Ascorbate

The Science of Vitamin C

Ascorbate

The Science of Vitamin C

Steve Hickey PhD.

and

Hilary Roberts PhD.

ISBN 1-4116-0724-4

Copyright © 2004 by Steve Hickey, PhD.

All rights reserved. No part of this publication may be reproduced, stored in a retrieval system, or transmitted in any form or by any means, electronic, mechanical, photocopying or otherwise without the prior permission of the copyright owner.

Contents

Contents	5
Acknowledgements	7
Preface	9
A new approach to vitamin C	11
Science and scurvy	19
Scientific reliability	31
Social influences on science	41
The history of vitamin C	51
Chemistry and vitamin C	57
Evolution and deficiency	69
How much does a healthy person need?	75
Is vitamin C safe?	93
Biased experiments	99
Oxidation and illness	109
The ultimate antioxidant	117
Dynamic flow	127
Heart disease and stroke	133
Heart disease or scurvy?	145
Vitamin C and lysine	157
Antioxidants and heart disease	163
Infectious diseases	173
Cancer	191
Vitamin C as a treatment for cancer	197
Replication and refutation	211
Glossary of terms	217
Index	227
References	231

Acknowledgements

We are grateful to all the following, particularly Dr Robert Cathcart, who read an early draft and contributed detailed comments and helpful suggestions. Dr Abram Hoffer read parts of the book and provided comment and encouragement. Dr Mark Levine provided detailed background information on his experimental results, despite our disagreeing with some of his conclusions. Dr John Pemberton supplied additional background information on his experiments on scurvy with Sir Hans Krebs. We should also like to thank Owen Fonorow and the Vitamin C Foundation for discussion on the Pauling therapy and reports from their web site. Dr Ian Brighthope kindly provided information about his current views. Dr Aleksandra Niedzwiecki responded to our questions about her work with Dr Mathias Rath on the treatment of heart disease with vitamin C, lysine and proline. Dr Ulrike Beisiegel provided comment about her collaborative work with Mathias Rath. Dr Lyubka Tantcheva provided information about her vitamin C research. Dr Tom Levy provided feedback on his work with heart disease.

Stephanie Morgan read a preprint of the book and provided supportive guidance and comments. Michael Roberts, FRCS, read parts of the manuscript and provided valuable comments. Simon Hickey and Andrew Hickey read the manuscript and made helpful suggestions. Michael Mesham, Nicola Chappell and Rose Chappell also provided comments and helped with the manuscript.

In memory of Tom Roberts

Preface

This is a book about the science of vitamin C (ascorbate), with particular emphasis on its use for treating and preventing disease. Our aim is to help people understand the controversy that currently surrounds this vitamin.

A large number of publications advocate the use of vitamin C supplements for good health and, in higher doses, as a treatment for disease. In this book, however, it is not our intention to promote the use of vitamin C as a treatment for any particular condition. In the event that a reader wishes to try vitamin C as a treatment for some condition, we must suggest that they first consult a physician. Medical guidance is essential for the treatment of significant disease and, for example, it is necessary to have an accurate diagnosis in order to determine the appropriate treatment. While high doses of vitamin C are normally safe, there are exceptions in certain disease states and genetic conditions. There may also be dangers in treating some forms of cancer; reports suggest that in some patients it may kill the tumour too quickly, producing necrosis and toxicity. In particular, medically qualified practitioners should carry out the administration of intravenous sodium ascorbate.

Nutritional supplementation, on the other hand, is a matter of personal choice; doctors often have little training in this area. The purpose of this book is to allow readers to reach an informed judgement about the health claims made for vitamin C. When considering large dose supplements, it is wise to ensure that the person does not have one of the few contraindications, such as iron overload or kidney disease. While medical assistance is important in treating disease, our view on supplementation is that until doctors have carried out the necessary research, they would do well to resist pontificating about a person's right to good nutrition.

The Nobel Prize winning physicist, Richard Feynman, used to say that if you really understood something you could explain it in a simple way. Our aim in this book is to present a straightforward and balanced account of the complex actions and potential uses of vitamin C. In some parts, the biochemistry is complicated, although we have tried to make it as simple as possible without introducing error. We hope that intelligent members of the public will be able to read the book without too much effort, even if they have not had a scientific education. We have also taken account of the needs of medical practitioners, who may be interested in the subject. Although the arguments are sophisticated, we have tried to represent them as clearly and simply as possible. We assume that scientists who read the book will be able to fill in the finer detail for themselves.

We have tried to make a difficult subject readable, while maintaining a high degree of accuracy. Many references are included, which should make it easier for the reader to follow up specific points. In referring to research papers, we have sometimes used the first or main author's name to represent a research group, as using *et al* or "and colleagues" disrupts the flow. We hope the other members of these research groups will not be offended; we do not mean to imply that a collaborative effort is the result of one person's work. In most cases, we have referenced specific research papers but occasionally we have included only a representative example. We have largely limited the discussion of specific illnesses to heart disease, infections and cancer: the largest killers in the industrial world. The list of diseases that may be influenced by vitamin C supplementation or treatment is much larger.

The primary aim of this book is to provide insight into an area of nutrition where rigorous science is generally lacking. Vitamin C was controversial long before Linus Pauling started to promote its use. Despite this, we still await well-designed experiments to determine the biological properties of the vitamin. Several researchers have suggested to us that the reason for this is that the questions are not particularly interesting, or are unlikely to produce positive results. To these, we would point out that it is unscientific to assume the results of experiments before they have been performed. Others suggest that commercial, institutional and financial forces actively prevent such research, at the expense of a sick population. Some critics have gone as far as to describe the actions of these influences as genocide. Our aim is to present a wide-ranging evaluation of the facts, so that the reader may come to their own conclusions.

For those who might be wondering, we published this book through Lulu to avoid editorial delays, retain control of the text and because they have a great publishing model. Since books are not subject to a formal peer review process, readers who notice errors or omissions are invited to email us at "radicalascorbate@yahoo.com". We will acknowledge helpful contributions in later editions.

A new approach to vitamin C

**"Solving a problem simply means representing it
so as to make the solution transparent."
Herbert A. Simon**

For decades, a controversy has raged concerning the effectiveness of vitamin C, also known as ascorbate, in preventing and treating disease. Many eminent scientists and several courageous clinicians have claimed the vitamin has widespread use against diseases such as heart disease, cancer and serious infections. If these claims are true, thousands of premature deaths and much suffering could be prevented. At the very least, the claims merit serious scientific investigation but this has proved difficult.

Supporters of vitamin C have had their applications for research funding denied repeatedly, and have had to be content with carrying out small-scale research projects and case studies. Detractors, on the other hand, have been funded lavishly. Despite this, they have carried out experiments that were different in crucial respects from those they were intended to replicate. The biased design of such studies provides misleading evidence against the use of vitamin C. The medical establishment claims to be scientific; this is the root of its authority. In the case of vitamin C, however, scientific principles have been violated and the public, together with the majority of the medical profession, has been tricked into believing pseudo-science.

Much of the research into the therapeutic uses of vitamin C is flawed or subject to misinterpretation. Our examination of the vitamin C controversy takes a scientific viewpoint, and asks whether the evidence supports the claims. In reviewing the evidence from each of the opposing camps, we provide explanations for the conflicting ideas and inconsistent experimental results.

This book will help you evaluate the medical evidence. We explain how to think like a scientist, being both critical and open-minded at the same time. Science, in theory at least, depends on the sharing and validation of results. In contrast, institutionalised medicine shapes current medical opinion according to multiple forces, and scientific accuracy often suffers when pitted against financial muscle. As this introduction suggests, the vitamin C story is complex, with many interested parties. In this book, however, we go back to basics and examine the key evidence for claims that large doses of vitamin C can prevent or cure several of the major causes of death in the developed nations. Some of these diseases are introduced briefly below.

Heart Disease and Stroke

Large doses of vitamin C may form the basis of a cure for coronary heart disease and occlusive stroke. Occlusive stroke occurs when a blood vessel is blocked and part of the brain is starved of oxygen. Similar blockage of arteries that supply the heart muscle causes coronary thrombosis. Heart disease is a leading cause of death in industrial countries. Current treatments include drugs, surgery and transplantation. Attempts at prevention include the use of other drugs and recommendations to change our diets. However, despite extensive medical and governmental advice to reduce fat and cholesterol intake, eat more vegetables and take more exercise, the death rate continues at a high level. People in their middle years, who are apparently healthy, can succumb to a sudden and fatal heart attack.

The claims made for vitamin C include the following: that it could form the basis of a complete cure for heart disease, that coronary thrombosis could become a thing of the past, and that the blocked arteries of people with atherosclerosis could be cleared out, so that they could live a normal life. It has even been suggested that, with appropriate levels of supplementation, no one need fear a heart attack again. Such a claim sounds too good to be true, and indeed, many physicians would find it laughable. However, it is the evidence that matters, not what people think. We will examine these claims in detail. By explaining the contradictory results of different studies concerning the involvement of vitamin C in heart disease, this book may act as a guide to future research.

Infectious disease

Another big killer, both in the developing world and in industrialised societies, is infectious disease. This remains true, despite improvements to diet and hygiene, along with the introduction of antibiotics, which have led people to believe that infectious diseases were under control. The successes of the 19th and 20th centuries may have caused us to become complacent and to think that disease-causing agents would be of little consequence in the near future. However, emerging viruses, such as HIV, SARS and Ebola, along with multiple drug resistant bacteria (e.g. MDR-TB, MRSA), are forcing us to think again.

Harmful infections result in damage to body tissues, caused by modified molecules called free radicals. Some physicians claim that large doses of vitamin C can act as the ultimate biological antioxidant, neutralising the free radicals and quenching the infection. The action of free radicals is similar in a range of diseases, from the common cold through to the haemorrhagic fever Ebola, so they might all be suitable for treatment with vitamin C. The only difference is the dose required. Could vitamin C really be a powerful antibiotic, effective against both bacteria and viruses? Again, we examine the evidence in detail.

Cancer

A Nobel Prize winner has suggested that large doses of vitamin C and other supplements could halve the incidence of cancer. In addition, he proposed that half the cancers that did occur could be "cured"; that is, the number of patients surviving at least five years from the time of diagnosis would double. Taken together, the claim implies that large doses of vitamin C and other nutrients would help approximately 75% of all patients with cancer. On first sight, this argument appears far-fetched: if such a cure existed, surely the numerous and highly supported cancer foundations would have found it years ago? Once again, we must suspend judgement and examine the evidence.

Mechanism of action

Vitamin C is the major antioxidant in the diet. We will show why antioxidants are not just another alternative health fad but are a central feature of life on earth. Most people have a vague idea that antioxidants are healthy, which is why they are included in cosmetics and health foods. However, to understand their actions in health and disease, we need to look at the chemistry of the body.

Antioxidants like vitamin C are essential for life in a hostile environment and, in biological terms, this includes our modern world. Disease processes almost invariably involve free radical attack, which antioxidant defences can modify. Many popular books present an "antioxidant miracle", promising an end to aging and disease; this is not one of them. Whenever possible, medicine should be based on scientific evidence rather than miracles. However, the support for vitamin C is compelling. When we examine the science without prejudice, the implications are even more forceful.

In order to understand why vitamin C is so important, we need to cover a certain amount of biology and chemistry. Vitamin C is also known as *ascorbic acid*; we will describe it in the text as either vitamin C or ascorbate. While we aim to be as clear as possible in describing the biochemistry, a certain amount of jargon is necessary.

vitamin C (ascorbate)

Patients and physicians

One aim of this book is to allow both patients and physicians to understand what has been going on in vitamin C research. To treat patients effectively, a doctor needs to keep up to date with current medical thinking, which can be time consuming. There are too many medical reports for anyone to be able to read them all. Instead, physicians rely on information provided in reviews of the literature, government publications and in data provided by drug companies. A general physician does not expect to act as a research scientist and often is not qualified to do so. Physicians follow a distillation of the results of medical research, known as "current medical opinion", which leads to ideas of "best practice". Deviation from this approach has to be justified. Although the aim of such conservatism is to serve the interests of the patient, it can limit the progress of medical science.

A patient visiting a physician expects to get clear, unbiased information about disease and treatment. Patients would like to have the information necessary to make informed choices about their treatment. In many cases, this does not happen.[1] For example, the physician may take the view that the patient does not have the intellectual capacity to understand the subtleties of the conflicting medical information. This is not as high-handed as it may sound: members of the public, who do not understand medical decision-making, often want a simple explanation.

The opposite view, assumed here, is that many intelligent members of the public would like to be sure that their physician bases medical decisions on a firm foundation.

Some doctors may feel that they need to give their patients a feeling of certainty about the treatment. A patient who believes in the efficacy of his treatment is more likely to have a positive outcome. In some circumstances, it might even be considered unethical to explain the uncertainty of outcome to the patient. The patient may need to be given hope. Physicians have been known to give chemotherapy to cancer patients they believed to be untreatable, since the patient or the family may demand any treatment that offers the slightest possibility of recovery. It can be hard for a physician to admit fallibility and to tell the patient that he can do no more. Dr Charles Moertel, a leading cancer specialist who performed some of the more controversial clinical trials with vitamin C, took this approach. He is on record as stating that the side effects of cytotoxic chemotherapy are worth suffering, in order to provide hope for the patient and to retain the doctor's authority.[2] He added that such an approach might also be useful in providing data for future treatments. However, it is not an option many people would knowingly choose as a patient.

The medical establishment

Medicine is not a scientific activity; it is a social, political and technological process for treating disease and other maladies. Its modern bias is on the treatment rather than the prevention of disease, and this emphasis may be related to the way it is structured. The "medical establishment" described here is not your local physician but the international organisation of medicine. Physicians generally care about their patients and do their utmost to provide treatment and support; they deserve respect for their efforts in what can be a difficult and demanding profession.

A rule-of-thumb for considering institutionalised bias is "follow the money". This is a well-known quotation of unknown origin, which serves as the primary investigative rule in understanding corruption. It is easy to see that the international drug industries, taken as a group, are a powerful influence on medicine and have strong financial interests. Governments are also influential: they foot the bill for much treatment, and receive taxes and donations from the pharmaceutical giants. In addition, they have a responsibility to regulate medicine in the best interest of the people and to protect us from quacks and charlatans.

Within medicine, there are specialised groupings that function rather like large industries. We can talk of a cancer industry,[3] which includes all those who have an interest, whether financial, power based, or career oriented, in the treatment and prevention of cancer. In addition, medical societies look after their members in a particular discipline, acting as a focus for deciding best practice and currently accepted opinion. Each large medical institution has its own agenda in protecting its image, organisation, and importance.

Medical bias

The bias against vitamin C reaches deep into the psychology of the medical establishment. To take an example, Norman Cousins' best selling book "Anatomy of an Illness" describes his experience of ankylosing spondylitis, a serious connective tissue disease that affects the spine.[4] His doctors estimated his chance of a full recovery at one in five hundred. Cousins was not ready to give up the fight, and devised a regime of laughter and large doses of vitamin C, up to 25 grams per day, for his treatment. Amazingly, he recovered. The medical establishment was quick to claim this was a clear case of the clinical benefit of positive thinking and outlook. Several accounts claimed this was a triumph for the placebo effect and holistic healing. Few remembered or reported the large doses of vitamin C. It seems that the climate of opinion is such that faith healing is considered to be on more solid ground than treatment with vitamin C.

The medical establishment has created an illusion regarding vitamin C, namely that large doses of the vitamin are of little medical interest and have been repeatedly discredited. In line with this, promoters of vitamin C are represented as unscientific, misguided quacks, advertising the substance for financial gain or publicity. People who believe that vitamin C and other antioxidants will do them good have been characterised as foolish or delusional. The establishment view has been that vitamin C supplements are a waste of money.

Despite this negative publicity, people all over the world have been supplementing their diet with vitamin C and other antioxidants, and the health food industries are becoming larger and more generally accepted. Offsetting this trend, vitamin C scare stories appear frequently in the media. As we shall see, claims for the benefits of vitamin C are often by leading scientists and clinicians with little to gain, whereas the same is not necessarily true for its detractors. The illusion that vitamin C has been discredited becomes less convincing as the scientific evidence is examined.

In the remainder of this book, we reflect only tangentially on political, organisational, or financial considerations, as they are secondary to scientific understanding. Those looking for a conspiracy by the medical establishment will be disappointed: there are no dark men in closed rooms deciding the fate of millions.

The medical establishment is an international community of conflicting and consistent aims and agendas. This community produces an environment of agreed information, with a bias towards conservative and financial considerations. Such an environment is probably natural and necessary for medicine to be a cohesive and reliable enterprise. However, peer-pressure can have powerful effects on members of groups. It is an established psychological fact that it is hard for an individual to avoid conformity to group ideals. When your career and reputation are important to you and depend upon group acceptance, iconoclasm is not an option.

Understanding the controversy

Vitamin C may prove to be an effective treatment for the major killers of western society. In this book, we will investigate some of the claims for vitamin C, filtering out weak arguments and pseudo-science, in order to gain insight into the value of this substance. As we shall see, vitamin C was controversial even before it had been discovered and isolated.

One reason why vitamin C therapy has been so controversial is that contradictory results are obtained with its use. Quite naturally, the negative results are taken as evidence that it has little clinical importance. Positive results, in contrast, suggest that it could be of great clinical value. We show that the observed inconsistency in the results of vitamin C is predictable, given an understanding of its method of action. Vitamin C does not act in the same way as a typical drug, or as a micronutrient. The confusion that surrounds vitamin C stems from a lack of knowledge about the way it works within the body. In particular, we highlight the importance of dosage: if too low or infrequent a dose is used, then little benefit is expected.

With the help of a simple unified model, we will describe the way vitamin C acts in disease states. This will allow us to explain how it could be an effective treatment for diseases as varied as cancer, infection and coronary heart disease. Studies that report no effect with vitamin C are also consistent with this model, as well as with the known biochemistry and pharmacology of the vitamin.

Science and scurvy

**"Never put your trust in anything but your own intellect ... always think for yourself."
Linus Pauling**

In this book, we take a scientific perspective on claims that vitamin C has unique health giving properties. To do this, we need to know how to think like a scientist, balancing the evidence in a way that is both critical and open-minded. Science is based on the sharing and validation of experimental results. The aim of this chapter is to help you evaluate scientific and medical research for yourself.

The scientific method

For a subject to be considered a true science, it must be based on the experimental method. Before we describe this, we will take a few lines to explain the difference between science and other bodies of knowledge. Technologies, such as electronics, are practical subjects that generally base their knowledge on scientific findings, but are not themselves classed as science. Despite the popular view, medicine, computer science and engineering are more properly viewed as technologies, although they certainly have components based on either the scientific method or mathematics.

Interestingly, mathematics is not a science, as its core methodology is a process of theorem, proof and corollary. A theorem is an idea that may have some evidence for its validity, but is not fully accepted until it is proven by a series of logical steps, starting from initial facts called axioms. Ultimately, proof means that mathematicians cannot conceive of any alternative. A corollary is something that follows immediately from the proven result.

Unlike mathematics, real science does not work according to the idea of "proof". It progresses by a related process of formulating a *hypothesis* (a testable idea), carrying out an *experiment* (a test of the idea), *replicating* the experiment (repeating tests) and *refuting the hypothesis* (showing the idea is wrong). To see how this works, we will describe one of the classic experiments in the history of medicine: James Lind's experiment on the treatment of scurvy.

A cure for scurvy

James Lind, the son of a merchant, was born in Edinburgh, Scotland. At the age of fifteen, he was apprenticed to a physician and, in 1739, passed the examination for Surgeon's Mate in the Royal Navy.

In those days, many sailors died of a disease called scurvy, now known to be caused by a prolonged deficiency of vitamin C. An adult whose diet contains no vitamin C will develop the disease in about six months. The symptoms are brutal: progressive body weakness, soft and spongy gums, and loose teeth. Blood vessels rupture and, in extreme cases, whole organs can seem to be mixed together, with an appearance similar to that of an invasive tumour. The loss of blood through ruptured vessels leads to severe anaemia. Scurvy has clinical effects similar to infectious diseases such as polio, or even the haemorrhagic fevers caused by emerging viruses such as the dreaded Ebola. Many of the symptoms of scurvy arise because the body needs vitamin C to help make its most abundant protein, collagen. Collagen is a large string-like molecule, used throughout the body to tie components together, for example in tendons and blood-vessel walls. It acts rather like the glass fibres in fibreglass composites. Without collagen, our bones would be weak and brittle, and we would collapse into an unstructured heap of cells.

Scurvy was of major medical importance in Lind's time, killing thousands of sailors and soldiers. On May 20, 1747, the 31-year-old Lind performed one of the earliest known medical experiments: he divided 12 scurvy sailors into six groups, with two subjects per group. All 12 men shared a common diet for breakfast, lunch and dinner but received a different supplement.

As can be seen from the table, the subjects receiving citrus fruit recovered fully. Both were well in six days. One subject returned to duty and the second was appointed as nurse to the subjects in the other groups. The two men who drank apple cider improved but were not well enough to work. None of the others showed any improvement. Something in the citrus fruit had cured the scurvy.

Lind's experiment was powerful. The power of an experimental result can be estimated using a measure known as *number-needed-to-treat*: this is the number of people that need to receive a treatment in order to save one life. If you need a large study to show an effect, say 10 cures in 10,000 subjects, then the effect is small and the number-needed-to-treat is high (10000 / 10 = 1000). There is minimal benefit of treatment in such a case, as you would need to treat a thousand patients to save a single life. In James Lind's experiment, the number-needed-to-treat scurvy with citrus fruit is roughly estimated at one (two treated / two recovered = 1): the most powerful effect possible.

Lind's experiment		
Group	**Supplement**	**Recovery**
Cider	Quart of cider per day	Slight
Vitriol	25 drops of elixir vitriol (sulphuric acid and aromatics)	None
Vinegar	Two spoonfuls of vinegar three times a day	None
Herb	Concoction of herbs and spices	None
Sea water	Half-pint of sea water daily	None
Citrus	Two oranges and one lemon daily	Full

Lind was not the first to suggest that citrus fruit could cure scurvy; the idea had been around for hundreds of years. It was noted in the medical literature in 1611, but even this medical reference lagged well behind unofficial recommendations. As long ago as 1227, Gilbertus de Aguilla had advised that sailors should carry a supply of fruit and vegetables on long sea journeys, to prevent scurvy. What Lind had done was to make the idea scientific by carrying out a controlled experiment, including an untreated comparison group. Because of this comparison group, Lind's study methods were a landmark in the history of medical science. We will evaluate Lind's test on dietary supplements to examine the important features of the experimental method. However, first we need to understand what an experiment is.

Ideas and theories

A *hypothesis* is an idea that can be tested. In James Lind's experiment, the hypothesis was that one or more of a range of dietary supplements would cure scurvy. A good hypothesis is an important feature of science and a major source of scientific creativity. Hypotheses are the ideas that bring new information into the scientific endeavour.

A good hypothesis has several recognisable features. Firstly, it is simple and can be stated clearly and without ambiguity. Secondly, it is testable, which means we can think of a practical test to prove the hypothesis is wrong. Notice how we aim to disprove the hypothesis, rather than trying to show that it is correct. Science does not prove things are true or attempt to show that a statement is a fact. A hypothesis gains validity when it is not shown to be wrong by experiment. When repeated tests fail to show it is incorrect, the hypothesis gains support and becomes an accepted theory. This leads to an inversion of the hypothesis, called a *null hypothesis*, for testing. If Lind's implied hypothesis was "citrus fruit cures scurvy", the null hypothesis was that "citrus fruit does not cure scurvy".

Lind's experiments were inconsistent with the null hypothesis that "citrus fruit does not cure scurvy", since the sailors taking citrus fruits got better. You will have noticed that the double negative of "disproving the null hypothesis" leads to tortuous language, so as long as we appreciate the practical difference between refuting a null hypothesis and accepting a hypothesis, we can ignore this complication. Incidentally, the effectiveness of vitamin C is frequently said to be "not proven" by the medical establishment. If we were being precise, we would say the phrase "not proven" is unscientific. The meaning of "not proven" in this context is that the person or group making this statement is not fully satisfied that the supporting evidence is conclusive.

A hypothesis, or a group of hypotheses, becomes a *theory* when it has withstood several experimental attempts to show it is wrong. A theory that is consistent with a body of established facts or experimental results, and makes predictions that have been validated, can be said to have scientific support. However, such support is fragile: given enough time, the theory is likely to be superseded. A good theory explains the way a part of the world works and, all things being equal, the simplest explanation is to be preferred. Theories are transient and become modified with time as more information is gathered. At some point, a new or competing theory is shown to be simpler or more accurate, and the old theory is abandoned.

With the accumulation of time and experiment, a simple rule that has not been refuted may become a *law*. A law is a part of a theory that is generally accepted as explaining some property of the universe with a degree of accuracy and simplicity. Newton's laws of motion and gravity were eventually superseded by Einstein's theory of relativity. We still use

Newton's laws, however, as they are simpler and easier to use: Einstein's relativistic error correction is too small to be important in most practical applications.

The power of refutation

There was a hypothesis that an early type of fish called a coelacanth had been extinct for about 65 million years. No-one had ever seen a coelacanth and its existence was known only from the fossil record. Despite the fact that no one had seen one over many centuries of fishing and exploring, it was not possible to say that it was definitely extinct. All that could be said was that the hypothesis that it was extinct was consistent with all the known facts. This idea was refuted in 1938, when a single coelacanth was caught off the coast of Africa and noticed by Marjorie Courtenay Latimer, the curator of a tiny museum in the port town of East London, northeast of Cape Town, South Africa. This is a clear example of a hypothesis that had a substantial amount of support being refuted by one observation. Even the largest amount of supporting evidence is weak compared to a single refutation.

Bias and misleading results

The results of medical and biological studies are often variable. Such variation is to be expected, but can make it hard to work out exactly what the study shows. One source of variation arises from the fact that each person is biologically unique and has an individual response to diseases and treatment.[5]

The complexity of biological processes makes the interpretation of experiments more difficult than with other disciplines. An experiment in physics, for example, can often be definitive. If you measure the speed of light, replications of the experiment should give the same result, within the level of accuracy of the procedure and conditions. Biological experiments deal with processes that are many orders of magnitude more complicated. If two experiments provided *exactly* the same answer, it would be a clear sign that something was wrong. Famously, the finding that a statistical measure of correlation did not change with the number of subjects led to the discovery that Sir Cyril Burt's work on IQ testing was fraudulent. As this case showed, even the most respected authorities can have hidden agendas. Biological experiments produce variable results and a clinical trial is rarely definitive.

Bias in presentation

There is a superb little book by Darrel Huff called "How to lie with statistics".[6] Huff's aim was to decrease innumeracy and show how it is possible to bias the presentation of hard numerical data. Often, in reading the literature on vitamin C, we have wondered whether various authors had read the book and were putting the ideas into practice.

One common way of misleading people is to present results as percentages.[1] An example might be the statement that "a new drug reduces the *risk* of getting a disease by 50%". This seems to imply a substantial (50%) reduction in the number of people getting the disease. However, this percentage has little relevance to the risk for a healthy person, because it conceals the prevalence of the disease in the general population. To take a concrete example, let us say a 50% reduction means that only five people get the disease instead of 10. Now, suppose there were 100 subjects in total, then, with the new drug, instead of 90 people being healthy, 95 people will be healthy. The improvement of five people in 100 is 5% of the total. However, if there were 10,000 people in the study, the improvement would be only five people in 10 thousand or 0.05% of the total. The "50% reduction in risk" translates as a 0.05% reduction in terms of a total population of 10,000 subjects.

By presenting data in terms of percentage reduction of risk, we can give a much more positive impression, as the previous example demonstrates. Not surprisingly, the effects of drugs are generally presented in terms of percentage improvement, which makes the results seem more impressive. To convey the correct meaning, results should be presented in terms of *absolute* reduction values. For example, the statement that "vitamin C will reduce the incidence of cancer from 25 in 100 to 12 in 100" provides clarity. *Relative* values, such as "vitamin C reduces the incidence of cancer by 50%", are misleading, because the reader probably does not know the initial incidence of the cancer.

Controlling for bias in experiments

An experiment is a test of a hypothesis. A good hypothesis is an idea that will suggest a critical experiment to demonstrate conclusively whether it is wrong. In the case of Lind's experiments, he selected six dietary supplements that might provide a treatment for the disease of scurvy, and gave them to 12 scurvy sailors. If his hypothesis had been wrong, the sailors would not have recovered. To be certain, he needed something to indicate that they had improved and he used a practical measure: their ability to return to work.

A feature of modern medical experiments is the *control group*. A control group is a group of subjects, as similar as possible to the test group, that are used to control for other factors that might influence the experimental results. The subjects in the control group might be given no treatment, a dummy treatment, or an established treatment. In James Lind's experiment, he compared six different treatments, each providing a control for the other groups. Since the experiment involved measuring something that changed (degree of illness), the aim was to make sure that the change did not arise from some other factor. The control group, as well as being similar to the test group, is subject to the same environmental and test conditions. So, if scurvy were caused by another factor, say sunlight, then by treating all subjects the same with respect to sunlight, we would expect no difference between the groups. Furthermore, if the control group is the same as the test group in terms of age, sex, ethnicity and other factors, the experiment may give results that are more sensitive.

The control group is a way of managing physiological, environmental and time related factors in experiments. However, there is another source of potential error, which stems from the fact that human beings are susceptible to suggestion: if they think they are being given a treatment, then they are likely to get better. This has been demonstrated by giving sick patients a dummy, or *placebo,* treatment, whereupon they are found to improve. We cannot think about animal or human experiments without considering the placebo effect. It has even been claimed that the history of medical treatment is largely the history of the placebo effect.[7] The placebo is often a sugar pill but it does not have to be an actual pill. A new x-ray treatment for terminal cancer patients, for example, has to prove itself better than just putting the person into the machine and not switching it on! In practice, a new treatment is often compared to an existing treatment that has already been shown to be superior to placebo. For the placebo control to work, it is obviously important that patients do not know whether they are in the placebo or treatment group: this is achieved by keeping them ignorant or "blind". An experiment in which the patients do not know which group they are in is called a blind study.

In addition to controlling for patient expectations, by means of placebos, another source of potential error is experimenter bias. The control for this is to design the experiment so that neither the researcher nor the patient knows to which group each patient belongs: this is called a "double-blind" study. The design of experiments has become standardised in the medical literature and the gold standard for testing new medical treatments is the "double-blind clinical trial".

A clinical trial is an experiment using test subjects suffering from a particular disease or condition. Subjects are allocated to either a treatment group or a control (or placebo) group. As mentioned above, a "blind" clinical trial is one in which the subjects do not know if they are getting the test treatment or a placebo. However, the placebo effect is strong and, if the physician knew which patients were in which group, he could behave differently towards the two groups, thus biasing the result by giving non-verbal cues to the patient. Because of the placebo effect, it can be important that even the physician or scientist conducting the experiment does not know which subjects are getting the new treatment. In a "double-blind" trial, neither the experimenter nor the patient knows who is getting the treatment, until the experiment is concluded and the results are analysed. To be even more certain that the experiment is not biased, patients are sometimes randomly assigned to the treatment or control groups.

Within medicine, the double-blind trial is given almost mythical status. However, it is important to remember that it is just a method for helping to minimise the placebo effect and experimenter bias. Lind's experiment was not even "single blind", as both he and the sailors knew who was receiving which treatment. Much current medical practice does not have a solid basis in double-blind clinical trials. Indeed, some areas of medicine are not suitable for placebo-controlled trials. Surgery, for example, is largely based on comparison with "gold standards" (procedures that are widely recognised as being the best available) and hence the foundation for robust evidence is lacking.[8] It is difficult to provide a placebo in surgery, as sham operations are not realistic or ethical. The best that can be achieved is a comparison with an existing technique. Ultimately, the validation is often based on historical evidence alone.

The ethics of double-blind clinical trials can cause problems. A new cancer treatment is likely to be compared with a current treatment, so that the patients are not refused treatment for a life-threatening illness. However, if the doctor is convinced of the greater efficacy of the new treatment and the disease is serious, it may not be possible to conduct this type of experiment. It is not ethical to give a placebo instead of an effective treatment for a serious disease. Who could sanction an experiment that might cost the lives of half of the subjects? In such cases, there are alternative methods of experimental design and analysis that can provide robust solutions.

The Case Study Method

One widely used method is the basic *observational report*, which is useful when studying factors that are difficult to quantify. In medicine, such reports are often called *case studies*. A case study is a report on one or more patients who have some interesting anatomy, disease or response to treatment. With certain rare diseases, the literature may be limited to case studies, as so few people have the condition that studying a group of patients is impossible. Case studies provide a method by which the early results of a treatment may be disseminated. Large-scale double-blind clinical trials are generally expensive and require resources that may not be available to a physician, whereas case studies can be reported wherever patients are treated.

The observational method has been responsible for many of medicine's greatest discoveries. As most schoolchildren are aware, simple observation led to the discovery of penicillin by the quiet Scotsman, Alexander Fleming. In 1929, Fleming found that a culture of the mould Penicillium inhibited the growth of both staphylococcus and streptococcus. These bacteria cause several common and serious diseases. The mould could be eaten and was apparently harmless to white blood cells in culture. Both mice and rabbits could tolerate injections of the mould. Fleming, however, did not take the obvious next step and show that an injection of penicillin could prevent or treat disease: Ernst Chain and Howard Florey carried out this experiment, about 11 years later. The spectacular success of the experiment led Florey and Chain to extract enough penicillin for initial clinical trials.

In the case of penicillin, a simple set of observations led to a small animal trial and directly to production of the antibiotic. Fleming's initial observations were critical to the process, but the follow up work of Florey and Chain was essential to produce a treatment. The development of antibiotics would have been slowed considerably if Florey and Chain had argued that Fleming's result was merely an observation and did not "prove" anything. Indeed, Florey's initial experiments on infected mice might themselves have been considered insubstantial, except for the fact that the treated mice lived and this result was both unprecedented and exceptional.

The penicillin story illustrates how, in certain circumstances, case studies or uncontrolled experiments can have equivalent or greater validity than double-blind clinical trials. Such a statement might shock many medical researchers, but it is true. Suppose it is an established fact that patients with a certain cancer all die within two months of diagnosis. In addition, let us assume that there are no reports in the literature of any patients ever having lived longer than two months, in any treatment trial, regardless of methodology used. If a study were then to show that all the patients getting a new drug lived for more than ten years, the result would be astounding. To object that the study is not double-blind and that use of the treatment should be delayed for several years until such tests had been performed, would be ridiculous. Some factor in the study is definitively producing an exceptional and unparalleled beneficial effect. A statistical anomaly of this magnitude could not easily be explained away as a placebo or environmental effect.

An outstanding observational result of this type should be replicated immediately. If the new treatment had a high safety margin and low cost, then it could be made available to patients even before knowing the results of follow-up studies, without medical, scientific or ethical objections. The development of penicillin proceeded in just this way. Situations such as this example have been studied in game theory for many years, and the optimal strategy is clear - it is a simple cost benefit analysis.[9] If the benefits (ten years life versus two months) outweigh the costs (cheap treatment, no known side effects) then the treatment has obvious value.

Results, replication and meaning

In his experiment, Lind used only 12 subjects and applied six different treatments, so he only had two subjects on whom to test each treatment. It is clear from the reported results that the experiment was of great consequence. A large result was observed in a small group, indicating the possibility of a strong effect. Conversely, if you need a large group to demonstrate an effect, the result is probably less important, as it affects a small proportion of subjects or the consequence is weak. It is easy to see that an experiment with ten subjects is unlikely to give a positive result for a treatment that will cure only one patient in one hundred. With a small group of subjects, you will generally find only the larger effects, although a fluke result might mislead you. In Lind's experiment, it was possible that the two people on citrus fruits were recovering anyway and he picked them by chance or unconscious bias. Thus, Lind could have tricked himself into believing a biased result. As the emphasis on double-blind studies suggests, an unbiased scientific experiment generally requires an impartial researcher.

It is possible to argue that James Lind's experimental results were just pure chance. They may have been due to a placebo effect, since the subjects knew the treatments and could have had expectations of them. Alternatively, the results could have been influenced by experimenter bias, if Lind had unconsciously conveyed his expectations to the sailors. Since only two subjects recovered, the conclusions rest on a small and possibly unreliable sample. Furthermore, Lind could not prove that he had not selected two subjects who were in the process of recovering anyway. It is also possible that the other subjects were also going to recover but the duration of the experiment was too short. More seriously, Lind could have purposely chosen two subjects who were recovering, to rig the experiment. The experiment could have been a complete fraud. Since at that time citrus fruits were relatively expensive, Lind could have had a vested interest, such as shares in "The Limey Company Limited - purveyors of fresh fruit". The long-term effects of eating citrus fruit could also have given cause for concern - what if they were later shown to cause cancer?

Lind's experiment was crucial because it set the stage for future investigation, by demonstrating that citrus fruit could apparently cure scurvy. Despite this, we have seen that it is possible to pull the experiment to pieces and suggest that it *proved* absolutely nothing. Such criticism of the experiment can be ignored for the following reasons. Firstly, it is the custom in science to assume that the experimenter is honest. Secondly, such criticism is inappropriate because *the experiment could easily be replicated.*

Replication is one of the fundamental tools of science. If you did not believe Lind's results, all you would need to repeat the experiment was a handful of citrus fruits and a few scurvy sailors, who were not difficult to find in the 1750's. You would get the answer in a week. If your results did not agree with Lind's findings, you could publish and explain where Lind went wrong. Almost anybody with an interest in the subject had the option to replicate the experiment and verify the results for themselves. Notice how the low cost and ease of replication makes the experiment more powerful.

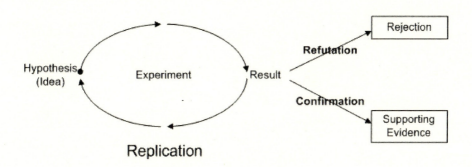

Lind had presented a strong hypothesis: that scurvy could be cured in a week with citrus fruits. He had reported a simple and persuasive experiment, and his conclusions were supported by the effectiveness of his treatment for scurvy. Lind, with his simple experiment, had followed the scientific method and his results were valid, unless refuted in the proper scientific way.

What followed Lind's discovery was rather typical of the history of medical science. His results, though interesting, were ignored. A year after his experiment, he retired from the Navy, obtained a medical degree and entered private practice. Ten years later, he was physician at the Royal Naval Hospital at Portsmouth. His recommendations to the British Admiralty, on feeding citrus fruit to sailors to prevent scurvy, were rejected. Citrus fruit was expensive and did not keep well. One person who did take heed of Lind's ideas was Captain Cook, the famous explorer, who decided to use stored vegetables on his voyages and found that no sailors suffered scurvy.

Lind's recommendations were eventually accepted by the admiralty in 1795, nearly 48 years later and one year after Lind's death. With hindsight, we can see that by ignoring the basic science for almost half a century, the British Admiralty were responsible for continued illness and a large number of deaths. After 250 years of observation and experiment, everyone now accepts James Lind's theory that something in citrus fruits will prevent scurvy, so it is easy to criticise the Admiralty, whose members were operating in the mist of conflicting ideas and arguments. Nonetheless, the Admiralty was responsible for the health of the sailors and could have repeated Lind's experiment at little cost. By ignoring evidence of the benefits of vitamin C, the modern medical establishment could be repeating the British Admiralty's failure to protect the health of its charges.

Scientific reliability

"Once you have learned how to ask relevant and appropriate questions, you have learned how to learn and no one can keep you from learning whatever you want or need to know."
Neil Postman and Charles Weingartner

Vitamin C has become controversial and it is therefore important to evaluate the evidence around it. To this end, we compare the characteristics of reliable science with that which is untrustworthy, and show how each can be recognised. Although the evaluation of scientific results is sophisticated, it is based on simple principles.

When it comes to designing and interpreting medical research, there is an urgent need to get back to the core rules of biology, chemistry and physics. Medical science is a branch of applied biology. Medical researchers do not have any special right to exclude evaluation of their work or ideas by other scientists, although some have tried to claim such rights in the past. An additional problem is that the primary purpose of a medical education is to prepare professionals to treat the sick, and scientific training is secondary to this aim. As a result, the established methods of science can be forgotten in the design and interpretation of medical experiments and clinical studies. Nevertheless, whether they like it or not, if the medical establishment get their science wrong, then their ideas will have to change.

Scientific fallibility

Although the scientific method generally works well, it can be fallible in the short term. An erroneous experimental result could lead people to think they had made a breakthrough. Strangely, the first replications might confirm the result, even though it is later shown to be invalid. With time, however, further investigation and experiment shows that the result was wrong and science moves on.

In spite of this, people sometimes hold on to erroneous results for many years, arguing that they are true. For example, the report of cold fusion, championed by Martin Fleishman and Stanley Pons, recently went through this process in public. Some independent laboratories initially confirmed their experimental results, but mounting experimental evidence conflicted with the theory, which most scientists now consider discredited. There are many similar stories throughout the history of science, including that of René Blondlots (1849-1930), the French physicist who claimed to have discovered a new form of radiation called N-rays. Dozens of other scientists confirmed the existence of N-rays in their own laboratories. Despite this, N-rays did not exist - they were later shown to be an artefact of the experimental procedure.

More recently, in 1966, Soviet scientists described a new form of water, called polywater. This anomalous water had a density higher than normal water, a viscosity 15 times greater, a boiling point higher than 100 degrees Centigrade and a freezing point lower than zero degrees. Over the next few years, hundreds of papers appeared in the scientific literature describing the properties of polywater. Theorists developed models, supported by some experimental measurements, in which strong hydrogen bonds were causing water to polymerise. After much study, it is now generally accepted that polywater was ordinary water that had become contaminated. The desire to believe in a new phenomenon can sometimes overpower the demand for solid, well-controlled evidence but, in the long term, both the N-ray and polywater hypotheses were refuted. These stories are a sign of healthy science and are to be expected. Nonetheless, in the cases of both N-rays and polywater, a surprisingly large number of respected scientists reported entities that did not actually exist.

Recognising suspect science

In 1953, Nobel Laureate (Chemistry, 1932) Dr Irving M. Langmuir (1881-1957), gave a colloquium at The Knolls Research Laboratory entitled "The science of things that aren't so". The talk was recorded, re-recorded and eventually transcribed and edited by RN Hall in 1968. In it, Langmuir presented a series of rules by which what he called "pathological science" could be recognised. These rules were intended to act as a set of indicators for recognising invalid hypotheses and experimental results.

As Langmuir illustrated in his talk, the desire to believe in a new phenomenon, such as N-rays or polywater, can be misleading. Fortunately, scientific knowledge is not absolute but is a continuing approximation, based on a self-regulating process.

Langmuir's rules:

- A causative agent of barely detectable intensity produces the maximum observable effect; the magnitude of the effect is largely independent of the intensity of the cause.

- The magnitude of the effect is small and close to the limit of detection; many experiments, measurements or subjects are needed to give statistical significance to the results.

- Great accuracy is claimed.

- Strange theories inconsistent with established scientific models are proposed, especially of a fantastical nature.

- Criticism is met by *ad hoc* excuses.

- The ratio of supporters to critics rises to a proportion of about 50% and then fades gradually.

A quick look at Langmuir's rules suggests that homeopathy and parapsychology are on dangerous ground. Perhaps more surprisingly, apparently reputable large-scale clinical trials, such as those performed on small amounts of vitamin C and other antioxidants, are also suspect, according Langmuir's second rule. In such studies, the magnitude of the effect expected is small and many subjects are recruited in order to give statistical validity. We are in the age of the mega clinical trial, in which thousands of subjects are followed for a number of years. In the case of vitamin studies, these trials are supposed to reveal relatively small effects of low doses of micronutrients. In such studies, a number of nutrients and effects are sometimes measured or estimated, before statistical methods are used to separate the effects of individual factors.

This large-scale approach is often an indicator of anticipated failure: if the experimenters expect the magnitude of an effect to be small, they will need a large number of subjects to demonstrate a significant difference between the experimental and control groups. Pharmacologist Professor David Horrobin states the position clearly: "If a trial has to be large, say more than 100 patients, it is large only because the expected effect size is very small". By this reasoning, large-scale studies need justification rather than admiration. Horrobin has gone so far as to suggest that large-scale trials on patients with rapidly fatal diseases are unethical.[10]

The starting hypothesis for such large-scale trials is often flimsy. For example, it may take the form that a small dose (a few milligrams) of a micronutrient, which is normally present in the diet in variable amounts, may be weakly associated (meaning it is hard to demonstrate the degree of causation) with some aspect of a chronic or long-term disease (meaning that it is difficult to show change). If the area is surrounded by ignorance, then scientists are forced to measure a number of factors that may be connected with the disease in some way, using a scattergun approach rather than a targeted shot. By contrast, when we understand a biological process, we can design a simple experiment to show a clear effect of a particular substance.

Large-scale double-blind mega-studies are the current fashion in medicine. The assumption is made that by increasing the number of subjects, the work somehow becomes more valid. Regrettably, these studies are borderline science, as a large-scale study is difficult to repeat and replication is at the core of the scientific method. Of particular concern are those studies that are so expensive that only pharmaceutical companies are ever likely to perform them. The way studies are financed and the potential source of any experimental bias or error become more important with larger studies, as they are more difficult to replicate. Replication reduces the possibility of bias. A large-scale study giving positive results is less convincing than, say, three smaller, equally positive studies by independent researchers, in different institutions, separately funded and using dissimilar methods. Despite our criticism, large-scale population studies have had some limited successes. For example, together with laboratory studies and clinical results, they helped to demonstrate that smoking leads to an increased risk of disease.

Most large-scale studies of vitamin C have involved small doses, around 300 milligrams or less. The expected effect of such a low dose is weak, making it necessary to have a large number of subjects. Governments and the press have incorrectly treated results from such studies as being of great merit, and findings are frequently applied to larger doses. For example, if a dose of 100mg of vitamin C is shown to have no effect, the same is taken to apply to a dose of 10 grams, which is 100 times the amount. Such extrapolation is unjustified. The large-scale small dose clinical trial is not a sensible way of looking for the substantial effects claimed for megadoses of vitamin C. The design of large-scale small dose trials takes us too close to Langmuir's rules, tending to produce unclear results that are difficult to replicate and will not be applicable to larger doses of the vitamin.

If you think you have a breakthrough cure for heart disease, you would be well advised to perform a small, simple experiment. Get a positive result with a few small experiments and there is a good chance you are on to something real. The small study approach favours treatments with low numbers-needed-to-treat. Small studies are useful for finding big, frequently occurring effects. They will generally filter out less common results, although there may be occasional anomalies due to chance. A small experiment showing a large effect, which independent researchers replicate several times, provides powerful evidence. Sadly, this forceful type of experiment has been avoided and denigrated by the medical establishment.

Langmuir's rules were intended to help keep ludicrous ideas out of medical and other sciences; however, the rules can be applied equally to established concepts and practice. We can use them to view the plausibility of current scientific, medical or nutritional practice, for example. According to the rules, large-scale clinical studies applying small amounts of vitamin C or drugs to show small positive or negative results should be regarded with suspicion. In any event, such studies have little relevance to our quest: we are interested in big effects, where the number-needed-to-treat is low.

Proselytisers and sceptics

It is essential to have an open mind when examining scientific results. We need to be wary of "proselytisers" who have a clear bias towards promoting vitamin C and provide only positive information. A proselytiser will only mention negative results in order to rubbish them. This group often promote more than one wacky idea at a time and believe in conspiracies, thinking that establishment groups are working together to suppress ideas. The promoters of a therapy can be misleading, but are ultimately less damaging to an idea than are extreme sceptics. Given time, the scientific evidence will bury any idea that is promoted but wrong.

Extreme sceptics are exactly the opposite of proselytisers; they claim to be scientific but are not. Sceptics claim that the new idea conflicts with established scientific fact and for this reason, they overlook positive results except when attempting to discredit them. The book that fuelled the vitamin C controversy, Linus Pauling's "Vitamin C and the Common Cold", was written specifically to answer the negative response of a well-known "sceptic", Dr Victor Herbert. Herbert wrote to Pauling, suggesting that he should desist from making statements about vitamin C for which there was no evidence. When Pauling wrote back with a list of scientific studies, Herbert rejected them as having no value. Herbert's response stimulated Pauling to write his book, to draw attention to the mounting evidence for vitamin C.[11]

While proselytisers talk of "conspiracies", sceptics talk of "charlatans" and "quacks". An important tool of the sceptic is censorship. Often the proselytisers have a financial interest in the sale of vitamins, but there are also professional sceptics. Some sceptics are paid by drug companies and other interested parties to deny findings that threaten profits and if this can be done by an apparently independent party, so much the better. Drug firms hire ghost-writers to write medical articles, supposedly produced by academics and doctors. The 'authors' of the articles may know nothing about the work but are paid handsomely for lending their names and reputations, and also benefit from extending their list of publications. It is estimated that ghost-written articles may account for almost half of the output of certain medical journals.[12] This illustrates how the pharmaceutical industry can use its financial muscle to pervert normal scientific controls.

The characteristics of a good theory

Langmuir's rules provide an insight into poor scientific theories and are useful as a guide to weeding out theories that are unworthy of investigation. For our purposes though, we need something stronger than a check that ideas are better than pseudo-science. We want to be able to show that the ideas are important, fruitful and likely to be of value. Fortunately, the characteristics of a good scientific theory can also be stated. The philosopher William Newton-Smith studied the rules for a good theory.[13] According to Newton-Smith, a good theory should conform to the following conditions (we have simplified the language).

Newton-Smith's rules

Good scientific theories should:

- Explain the success of previous theories.
- Be fertile and generate new ideas for future research.
- Have a history of providing correct predictions.
- Provide additional support for established or related theories.
- Need few additional hypotheses to explain failures.
- Be internally consistent and not contain contradictions. Be consistent with other scientific models.
- Be as simple as possible.

As we progress, we will describe a theory of action for vitamin C that is consistent with these rules. This approach will include previous ideas and take into account the history of scientific studies of the action of large doses of vitamin C. We have based this theory on existing ideas of free radical chemistry in biology and medicine. The theory of action developed is internally consistent and explains why varying results are often obtained in experimental studies. We do not require any additional hypotheses to explain the available experimental results. The model can be applied to numerous disease processes and makes strong predictions, providing simple, testable hypotheses.

In later chapters, we look into the substantial effects reported for large doses of vitamin C (several grams and above). The downside to looking at large doses is that clinical studies are in short supply. It is easy to find studies of the action of small doses of the vitamin, but these often give variable and inapplicable results. One theory of the pharmacological action of vitamin C, proposed by Dr Robert Cathcart,[14] suggests that its clinical efficacy is sharply dose dependent and has a threshold. A study of 100 milligrams per day of vitamin C, even in one million subjects, gives essentially no information about the effect of a three-gram per day supplement. To take a simple analogy – if a quarter of one child-dose aspirin does not cure my headache, can I assume that three adult tablets will not work either?

Asking the right questions

One of the criteria that identify great science is the quality of the question being investigated: a good question is often more important than a good answer. A well-defined question will often suggest the form of the experiment needed to give the answer. A strong hypothesis is one that can be answered by a simple experiment giving unequivocal results. Vague questions lead to unclear experimental results and a precarious form of science.

The biology of vitamin C is complicated and one reason for the controversial nature of this substance is that people have started by asking the wrong questions. To clarify matters, we will now list some fundamental questions in the area of vitamin C and health.

Preventing Scurvy

The first important question is "What dose is required to prevent acute scurvy in a normal subject?" This question is easy to answer and is uncontroversial. Five to ten milligrams of vitamin C per day will prevent a person dying an unpleasant death from acute scurvy. James Lind effectively established the answer to this first question, but it has been confirmed many times over the last 250 years, both by direct observation and experimentally.

Good health

The second question is "How much vitamin C does a person need to maintain good health?" This question is clearly not the same as asking how much vitamin C would be needed to stop someone being very ill and dropping dead with scurvy in the next few months. Despite this difference between "not dying of acute scurvy" and "being in good health" being obvious when stated clearly, many doctors and nutritionists have argued that if you have enough vitamin C to stop you getting scurvy, you do not need any more. If evidence gathering is restricted to finding the amounts needed for prevention of acute scurvy, then that is all we know. A person who is getting 10 milligrams of vitamin C daily will not die of acute scurvy this year, but this does not mean they are receiving an optimal amount.

It may be that people with such a low vitamin C intake are, in the longer term, in danger from heart disease, cancer, infection, and other diseases. The extrapolation from "not getting scurvy" to "being in good health" is unjustifiable. A confused section of the medical profession has been arguing in this way ever since vitamin C was isolated decades ago. To find out how much vitamin C is needed for good health, we need a measurable definition of good health, otherwise the question is philosophical rather than scientific.

A healthy population

Thirdly, we want to know how much vitamin C is required to make sure *everyone in the population* is getting enough for good health. This is quite a different question to asking how much a single individual might need. Humans vary in more fundamental parameters than sex, shoe size, hair colour and age. Each individual is biologically unique; references to such variation in human biology are becoming everyday occurrences, with the forensic use of blood types, fingerprints, tissue typing and DNA testing. Similarly, each individual has specific requirements for vitamin C. Since it is possible to measure the variation between individuals, the recommended minimum vitamin C intake should be high enough to cover the whole population, including those people with a higher than average requirement.

Preventing disease

The next question we would like answered is, "What amount of vitamin C is necessary for disease prevention?" If the daily dose is enough for a varied population to have long-term health in the absence of disease, it does not mean that it is necessarily optimal for disease prevention. One of the postulated benefits of ascorbate is that it will help prevent colds and other viral diseases. If the amount needed for preventing disease is higher than the amount needed for good health, we may need to take more to avoid unnecessary illness. This area is on the borderline between nutrition, the amount we need to take to stay healthy, and pharmacology, when we use vitamin C as a drug for a direct biological effect.

Treating disease

A critical question concerns the amount of vitamin C required to treat an infectious or other disease once it has taken hold. This is a pharmacological dose and may bear little relationship to the amount required daily for good health. In the history of the use of vitamin C, it is apparent that there is considerable confusion about the difference between a pharmacological dose of the vitamin and a nutritional dose. This led to the use of inappropriately low doses of vitamin C in treating disease or in replicating experiments where positive findings had been reported. Forty grams of vitamin C is clearly a pharmacological dose, while 250 milligrams is a nutritional dose.

Toxicity

The question of toxic effects of vitamin C is relevant here. When the dose increases beyond the level necessary to prevent acute death from scurvy, we may need to balance the level of the dose against any potential toxic effects. Below this level, toxicity is not a serious question, as the effects of poisoning with vitamin C would have to be dire to be worse than dying from scurvy. We shall see that vitamin C is not only an essential requirement of the diet, but is very safe indeed. However, although the general safety of ascorbate is accepted, hypotheses of toxicity from very large doses are as valid as those concerning benefits and should be treated with the same rigor. We should remember that not long ago, even smoking was considered by many people to have health benefits.

Social Influences

In thinking about these questions, it is important to note that science as practiced is often biased. Science is a social activity and, like all human actions, can be flawed. We have stressed the dose dependent nature of claims for ascorbate as a treatment for various diseases. With this in mind, a sceptical scientist wanting to refute a positive experiment with vitamin C on, say, the common cold, could repeat the experiment with a sub-therapeutic dose of the vitamin. This would minimise the likelihood of a positive effect. In the following chapter, we consider the social pressures that have influenced the research on vitamin C as a therapeutic agent.

Social influences on science

"Science is the belief in the ignorance of the experts." Richard Feynman

There is substantial resistance from the medical establishment to the claims made for vitamin C and, in order to understand why, we must first look at the way that scientific knowledge develops. The suggested health benefits of antioxidants such as vitamins C and E are substantial, but they need to be verified, or at least be consistent with current scientific ideas, before they are accepted.

Science is a social activity and does not proceed by merely accumulating knowledge. Thomas Kuhn discussed how scientific ideas progress, in his book *The Structure of Scientific Revolutions,* published in 1962.[15] A scientific revolution, also called a *paradigm shift*, occurs when a new idea replaces an established model or theory. Scientific revolutions start when a group of scientists begins to sense that an existing theory has ceased to be the best explanation of the facts. A scientific revolution does not necessarily involve an earth shattering idea, but implies the replacement of a currently accepted model. To people who are not actively involved, it can seem to be just part of the process of scientific development. However, to the players in the field, it can be the maker of careers or the destroyer of reputations.

A scientific revolution can be traumatic. Kuhn suggests that large changes in science are similar to those in politics. While new ideas are in the process of being accepted, scientists are often divided into competing camps or groups. One group seeks to defend the status quo while the other tries to overthrow it. Political dialogue breaks down when a certain level of polarisation has occurred. Since the revolutionaries find themselves outside the current institutional structure, they resort to the techniques of mass persuasion.

The repeated rejection of new scientific theories is a historic fact, but does not automatically imply ignorance and confusion. By definition, a paradigm shift involves the rejection of an accepted scientific model. Science progresses when a new idea is substituted for an older, less useful interpretation of the data. Because a paradigm shift often involves social change, it is sometimes incorrectly claimed the argument cannot be resolved by the methods of normal science alone.[532] The two competing groups often present circular arguments in defence of their theory.

However solid the arguments, they do not provide a communication bridge to the other camp. The competing theories, and sometimes their entire frames of reference, are incompatible.

On occasion, the fact that ideas are currently accepted has been used to justify the establishment position. Such ideas are assumed tried and tested, and therefore more likely to be correct. This adds to the confusion that can occur between the accuracy of an idea and its degree of acceptance. Experience shows that current acceptance is no guarantee of correctness. Without doubt, many standard ideas in medical science will eventually be superseded by others that are more accurate.

Galileo Galilei and the Inquisition

Galileo's belief in the Copernican System, which states that the earth and planets circle the sun, got him into trouble with the Catholic Church. The Inquisition upheld the authority of the Church, with the aim of eradicating heresies. A committee of Inquisition consultants declared that the Copernican proposition was heresy. Cardinal Bellarmine warned Galileo that, by order of Pope Paul V, he should not discuss or defend Copernican ideas. Later, in 1624, Pope Urban VIII assured Galileo that he could write about Copernican theory as long as he treated it as a mathematical proposition rather than reality. However, when Galileo's book, "Dialogue Concerning the Two Chief World Systems" was printed, he was called to face the Inquisition. Galileo was found guilty of heresy and placed under house arrest in his home near Florence. Galileo had a tough time, but even the authority of the Catholic Church could not change reality. The earth circled the sun, whether the Church accepted it or not.

Patents and profits

Medicine is a highly regulated activity and much of the regulation is aimed, quite properly, at keeping patients safe from dangerous potions and treatments. Before a new drug is released, it must pass through a process of experimental and clinical validation. The cost of this validation has risen with time, hence the introduction of new drugs is now limited to those organisations that can afford the testing. Pharmaceutical companies invest in the development of new medicines and get their returns in the form of revenues from patented drugs. Once a drug has been patented, the company has an exclusive right to produce the drug and can charge a substantial premium for the duration of the patent. For large pharmaceutical companies, this situation is clearly advantageous, as it reduces the competition for their highly profitable patented drugs. By comparison, health food companies have relatively small turnovers and cannot afford to validate unpatentable nutrients.

The primary interest of a pharmaceutical company is profit for its shareholders. The company charges large fees for new drugs that it holds under patent. As in any good business, it actively promotes its own products relative to those of competitors. The high cost of new drugs is typically justified by the need for a continuing research and development programme to find and produce new and effective drugs, although the research budget is often a small fraction of the company's turnover. Provided it has some novel features, a new drug is patentable, whereas a simple nutrient, such as vitamin C, cannot be patented. However, simple low cost treatments, which could put patented medicines out of business, threaten the profits of the pharmaceutical industry.

To maximise profits from a drug, the company must influence both the medical organisations and the prescribing doctors. Many doctors would argue that this influence has been overstated: they see themselves as sophisticated professionals, far too perceptive to fall for drug company promotions. This is clearly an area of some controversy. The pharmaceutical industry spends vast sums funding research projects, not to mention conferences in exotic locations. Many eminent and lucrative medical careers are dependent on the success of particular products. Clearly, it is difficult to remain impartial under such pressure. An increase in scepticism in this area would be most welcome.

While active drug company competition can be good for patients, it can also lead to anomalies. Most research relates to diseases of the affluent, where substantial profits can be achieved. An African country, with high rates of malaria and HIV/AIDS, may not have the money to buy a new and expensive treatment that would greatly help its people. International intellectual property law prevents poor countries from manufacturing new drugs for their own people, at an affordable price. The distortion of ethics that occurs when the financial interests of large companies clash with health has been highlighted by the behaviour of the tobacco industry, but is also observed in the food and pharmaceutical fields. Incidentally, the medical establishment has little difficulty recognising the analogous conflict of interest that occurs when health food companies promote certain vitamins or nutritional supplements. Indeed, they have gone to such lengths to prevent it, that in many countries labelling information on supplements is severely restricted by law.

The relationship between high profit medical industries and the medical establishment is sometimes presented as revealing a hidden conspiracy. There is, however, nothing hidden and little is conspiratorial. The organisations are simply too large and powerful to care much about diffuse public criticism. The United States Food and Drug Administration (FDA) does not hide its association with the pharmaceutical industry. It openly receives funds from them for drug certification. This means that the FDA is financially dependent on the companies and subject to direct commercial pressure from them.[16] At the time of writing, companies pay about $300,000 to apply for approval of a new drug, as well as about $145,000 for each manufacturing establishment and an amount for each product. These payments are insignificant to large pharmaceutical companies, although they are big enough to ensure that health food suppliers are unable to register their products for financial reasons.

Even the British Medical Journal has become concerned about the inappropriate links between the FDA and the drug companies.[17] In a recent issue, the BMJ suggested that these links were detrimental and could allow drug companies to push through drugs that are largely worthless or even dangerous.[18] They used the example of the drug Alosetron, for the treatment of irritable bowel syndrome, which had been withdrawn because, although its benefits were slight, there were frequent side effects and some deaths. The FDA recommended that the drug be made available again, as long as the responsibility for the side effects was with the issuing doctor! These unfortunate results are only to be expected, given the great economic power and scale of the profit-making health industries.

To highlight the long-term influence of finance on medicine, we recall James Lind's experiment on scurvy. Previously, we suggested that the long delay in the use of citrus fruit by the British Admiralty was a result of institutional inertia and ignorance. The historian, J.J. Keevil, has proposed an alternative explanation for this delay, the press gang.[19] Keevil describes the situation eloquently: "When we recall that the naval manning problem arose principally through losses from disease [and] that much of this was scorbutic or intestinal, the failure to obtain fresh provisions at every opportunity can be accounted for only on the ground of economy. It was in fact cheaper to replace seamen."[20] Economics is a principle factor in delaying the social change associated with a paradigm shift in medicine or science. Just as it may once have been cheaper to let sailors die of scurvy and replace them using a press gang, it can be more profitable to promote a patented drug than to accept the benefits of nutritional medicine.

Snake-oil?

The establishment has likened people explaining the benefits of vitamin C to peddlers of snake-oil. This turns out to be an interesting analogy. Dr Udo Erasmus recently investigated the origin of the derisory term snake-oil.[21] He found that snake-oil was a traditional Chinese treatment, introduced to the United States by Chinese labourers laying the early railway lines. They persuaded fellow workers to use snake-oil, to reduce inflammation and arthritic pain. Rubbing snake-oil into the skin was said to bring symptomatic relief. The sellers of patent medicines saw the use of snake-oil as a threat and disparaged the use of this treatment, to the extent that snake-oil continues to have a bad name to the present day.

In 1989, a Californian physician, Dr Richard Kunin, investigated the properties of oil from the Chinese water snake.[22] This snake-oil, used in Chinese medicine, was found to contain the highest proportion of omega-3 fatty acids in a natural oil (20%). Since omega-3 fatty acids are potent inhibitors of inflammation, there is every reason to suppose that snake-oil was effective. The term "snake-oil" actually refers to the use of a now established anti-inflammatory substance, in a treatment suppressed for financial gain.

Safety

Physicians operate in an environment that is highly regulated by governments and related organisations. Large drug companies are dedicated to making sure that they have the latest and most lucrative patented medicines, but they do have to comply with safety regulations. Medicines in current use have been tested, either by a long period of medical practice or by resourceful drug companies, watched over by government inspectors. This regulatory system was designed, in part, to stop quacks from introducing new treatments that either do not work or are actually dangerous. However, the intended safety mechanism has overrun its original aim and can sometimes repress innovation in medicine, especially if it originates on the fringes or from outside the system.

Folic Acid

Many physicians have claimed that people get all the vitamins and nutrients they need from a balanced diet, and that there is no need to take supplements. To illustrate the ignorance and bias in this assertion, we need only look at one of the B vitamins, folic acid. In 1981, Professor

Richard Smithells showed that supplementing pregnant women with folic acid reduced the incidence of neural tube defects.[23,24] The neural tube is the early form of the spine, during embryonic development. A major form of clinical neural tube defect is spina bifida, in which the spine is not completely formed and the spinal cord protrudes from the lower back. People born with spina bifida are often crippled for life and may have further problems, such as incontinence. Supplementation with folic acid can reduce the incidence of spina bifida by 70%. Without supplementation, the incidence is about 25 cases for each 100,000 births, which is reduced to about eight cases if the mother takes folic acid supplements.[25]

The medical establishment was hostile to the claim for folic acid and did not accept the findings.[503] This hostility to nutritional treatments is a familiar feature of medicine. In the case of spina bifida, the link to the diet was established and had been known for centuries.[26] In the 18th century, the midwife Catherina Schrader of the Netherlands kept detailed records. In the years 1722-1723 and 1732-1733, she noted that large numbers of neural tube defects corresponded with crop failures. The defects occurred largely in the babies of poor urban women, with limited access to good food.

Smithells had also reported the connection as early as 1965.[27] Despite this, the establishment would not make a recommendation for folic acid supplementation, until further clinical trials had taken place and confirmed the results. It took an additional 11 years to provide sufficient supporting evidence. Presumably, the justification for the delay was based on an analogy with drugs, which are required to demonstrate effectiveness and safety, before their use can be recommended. However, folic acid is not a drug: it is a normal part of the diet. The refusal to recommend folic acid was a result of failure to carry out a simple cost benefit analysis. There was no significant danger associated with supplementation, which would have ensured that the mother was not deficient. The refusal to recommend folate meant that, for over a decade, babies were born with spina bifida unnecessarily. Actually, it is worse than this, since the general insistence that vitamin supplements were not necessary extended long before the studies on neural tube defects. Women not deterred from taking supplements would have had fewer damaged children.

It is easy with hindsight to criticise the doctors for getting it wrong with folic acid. The problem remains that they have been subject to criticism for inappropriate action on nutrients for decades, yet have not responded. If a deficiency of folic acid has relatively recently been found to produce spina bifida, what other problems might deficiency cause, that we do not yet know about? It has emerged that numbers of the related problem anencephaly, being born without a brain, are also reduced by supplementation with folate. Folic acid deficiency may cause other birth defects - we simply do not know. Then there may be birth defects caused by deficiency in others of the vitamin B complex, or indeed any of the vitamins.

Recent research suggests that marginal deficiency in vitamin B7, also called biotin or vitamin H, sometimes causes birth defects.[28] According to the establishment approach, each defect caused by such a deficiency would first have to be identified. Then there would be a delay of at least a decade, while the medical establishment performed clinical trials to confirm the relationship. At this point, they would issue a recommendation stating that supplementation with the vitamin is a good thing to do. The whole procedure is clearly ridiculous, as we are considering vitamin *deficiency*. The harmless and cheap recommendation to take a suitable multivitamin supplement, while the trials were being carried out, would prevent these problems.

A modern diet does not provide the whole range of essential nutrients. Most people do not eat a diet prescribed by a nutritionist. Perhaps the experts still suggesting that people get all the nutrients they need from a balanced diet should examine the real world: people eat junk food. More importantly, if such "experts" can make a mistake as fundamental as failing to understand that pregnant women should not be deficient in folic acid, why should we believe them on any aspect of nutrition? Medicine has a long history of serious mistakes that occurred because scientific results were ignored. With the benefit of hindsight, medicine at any point in history appears to contain several strange and barbaric practices that are defended with unjustified arrogance. It seems unlikely that our own age will prove to be exempt from such errors.

Child birth fever

It frequently takes many years for medicine to respond to new ideas, even when there is solid evidence. To take a historical example, Ignatz Semmelweis, a German-Hungarian physician, studied the high death rates in hospitals during childbirth. Maternal death rates from puerperal (childbirth) fever were high, especially on wards attended by physicians and medical students who carried out post-mortems.[29,31] Semmelweis thought the disease was caused by "odours" and this led him to suggest that physicians should wash their hands between patients and eliminate any smell using chlorine solution to prevent infection. In the month following the introduction of this practice, the mortality rate fell dramatically.

Despite these convincing findings, Semmelweis was not lauded as a medical hero. His suggestion ran contrary to the prevailing understanding of disease processes and offended the doctors' dignity. Evidence alone would not convince them. Even though popular opinion looked upon childbirth in hospital as almost a death sentence, the hospital authorities were unmoved. Semmelweis was dismissed from his position and had to return to his native Budapest. He was eventually offered a part-time post in an obstetric hospital, but only on condition that he would agree not to instruct medical students in washing their hands. Semmelweis was literally driven crazy by such reactions to his suggestion that good hygiene prevented infection. Rather than repeat the experiment and check the results, the medical establishment resorted to argument. By attacking the source rather than the idea, they could ignore the proposal and its implications for their understanding of disease.

Semmelweis was not the first to suggest that washing hands and other basic hygiene would prevent childbed fever. In 1795, Dr Alexander Gordon of Aberdeen, Scotland, made a similar suggestion. Dr Oliver Wendell Holmes, an esteemed Harvard professor, presented a paper in 1843 entitled "On the contagiousness of puerperal fever" in which he argued for hygiene and hand washing to prevent the disease.[30] These doctors were luckier than Semmelweis, as their suggestions were simply ignored.

Stomach ulcers caused by bacterial infection

The Semmelweis story may be dismissed by modern medical researchers as irrelevant to contemporary scientific medicine. However, recently the hypothesis that stomach ulcers were a result of bacterial infection was treated with disdain, simply because the evidence presented did not agree with the current model. Everyone "knew" that stomach ulcers were not caused by bacteria. Then, in 1983, Drs. Robin Warren and Barry Marshall found a bacterium, called Helicobacter pylori, in the stomachs of people with gastritis. They hypothesised that peptic ulcers are caused by bacterial infection rather than by excess acidity or stress, as was currently believed. At first, their hypothesis was considered preposterous. However, it is now accepted that infection plays an important contributory role in causing many ulcers. Stomach ulcers are common, affecting up to 10% of people, and the current evidence is that destroying the bacteria will cure many cases of this disease.

Do no harm

Medical experts often suggest that a high degree of conservatism is required in the acceptance and authorisation of new medical treatments. However, this rule applies more to some treatments than to others. A new treatment must be shown to be both effective and harm-free. In the case of new drugs, this statement is certainly correct, since many have significant side effects. On the other hand, the safety of vitamin C is already established, as it is a normal part of the diet and is even essential to life. Given that a treatment is known to be safe, the evidence required for its effectiveness is lower. For an extremely safe treatment, we only need to know that there is some small advantage to its use. Indeed, even if there were no therapeutic value, it could still have a beneficial effect as a placebo. These arguments are clear and are covered in detail by medical decision-making and game theory.[9] The degree of conservatism required with a new treatment depends on its known safety.

Sociology or science?

Thomas Kuhn's ideas are often misunderstood. He was talking about sociology and the acceptability of ideas by the current establishment. He explained how humans, especially in groups, are often resistant to new ideas. His ideas on paradigm shifts have little bearing on the scientific method itself. The concept of a paradigm shift is not based on the *validity* of a new theory but on its *acceptability* to the majority. Kuhn's argument that a paradigm shift dispute is political and cannot be resolved by the scientific method is only true in the short term. Both sociology and medical politics are secondary to science. Unfortunately, there is no rule for how long it will take for the new science to replace the old,[31] and with vitamin C, as in the past with Lind's recommendations for sailors, the delay may be measured in lives as well as years.

Surprisingly, the idea that a substance in citrus fruit could cure scurvy was questioned again in the early 20th century. As late as 1911, it was claimed that scurvy was caused by contaminated food, especially tainted meat. By the time of the First World War, confidence that citrus fruit and vegetables would prevent scurvy was low.[20] We can now explain the anomalies that led to this lack of confidence, as we know the chemical properties of ascorbate, its distribution in foodstuffs and its stability during cooking. Heated juice or stored citrus fruit did not necessarily prevent scurvy. People consuming uncooked meat or milk could survive for months without scurvy, despite eating no vegetables. Since vitamin C had not yet been isolated, it was not known that it could be destroyed by heating or oxidation, and outbreaks of scurvy still occurred. Nonetheless, thousands of lives had already been saved by including citrus fruit and other vegetables in the diet. If scurvy did occur, repeating Lind's experiment would have given unequivocal results.

Refutable knowledge

The true progress of science has been described as the acquisition of *refutable knowledge* and our understanding of this owes much to the work of Karl Popper,[32,33] the leading philosopher of science in the 20th century. Popper concerned himself with the generation of scientific knowledge. This involves inventing and testing theories. The key distinguishing feature of scientific knowledge is that it should be refutable. Popper does not describe how science is actually performed on a day-to-day basis, but he does provide a focus on the real world when considering scientific ideas.[34] Popper indicates that any attempt to place political or social constraints on science is inevitably disastrous. Any group, institution, or pattern of belief that tries to constrain science is harmful.

Separating science from sociology

Ultimately, there is only one rule in science, which is to describe the world accurately. For example, an aeroplane is designed according to scientific principles, with the requirement that it flies with a degree of controllability and safety. In building an aeroplane, the primary consideration is that it will do its job and function well. No amount of public relations by the airline company would compensate passengers for an aeroplane's inability to fly or, even worse, to land.

The situation in medicine is essentially the same. A scientific theory needs to be supported by the relevant experimental data. It does not matter what the eminent authorities argue or whether no one believes the theory. A theory that is consistent with the facts will eventually be accepted. The core scientific questions we should ask are these: How convincing is the evidence? Do the theories fit the facts? Is the evidence consistent? What do the experiments actually show? In medicine, despite the jargon, ideas are generally simple when explained clearly. Given the evidence, intelligent people are able to grasp the essential ideas and make up their own minds.

The history of vitamin C

"In questions of science, the authority of a thousand is not worth the humble reasoning of a single individual." Galileo Galilei

The value of certain foods in preventing illness was known long before the first vitamins were actually identified. We have already described Lind's experiments, which showed that citrus fruit would cure scurvy. Despite these observations, it was not until the 20th century that the existence of vitamins was established. In 1897, a Dutch physician called Christiaan Eijkman discovered that feeding fowl a diet of polished, rather than unpolished, rice caused a condition similar to the disease beriberi. At first, Eijkman did not recognise beriberi as a deficiency disease but, by 1907, he and his collaborator Grijns had concluded that rice bran contained a nutrient that is essential for good health. Eijkman's findings led to the discovery of vitamins and he was awarded the Nobel Prize in recognition of his work.

In 1906, the British biochemist Sir Frederick Hopkins showed that foods contained essential accessory factors. Hopkins fed rats on a diet of artificial milk, made from protein, fat, carbohydrates and mineral salts, and found that they did not grow. However, they grew rapidly when a little cow's milk was added to the formula. He termed the missing factors, now known as vitamins, *accessory substances*. Casimir Funk, a Polish biochemist, followed up Eijkman's work in 1912 by isolating a substance from unpolished rice that would prevent the disease beriberi. Funk demonstrated that pigeons could be cured of beriberi by feeding them a concentrate made from rice polishings. He discovered that the active substance was an amine (a type of nitrogen-containing compound), so he suggested the name vitamine, short for "vital amine". This term came to be applied more generally to what Hopkins had called "accessory factors", which became known as *vitamins* when it was realised that not all such substances contained nitrogen. We now know that vitamins have different chemical properties and functions, and many are not amines.

In 1912, Hopkins and Funk proposed the vitamin hypothesis of deficiency disease, which postulates that the absence of sufficient amounts of a particular substance in a system may lead to a corresponding disease. They suggested four "vitamines", providing protection against four diseases: beriberi (thiamine: vitamin B1), scurvy (ascorbic acid: vitamin C), pellagra (niacin: vitamin B3) and rickets (vitamin D). Part of the hypothesis stated that scurvy was a dietary deficiency disease caused by a lack of an unknown water-soluble substance, called vitamin C. Hopkins was awarded the 1929 Nobel Prize for Physiology or Medicine, for discovery of essential nutrient factors, now known as vitamins, needed in animal diets to maintain health. He received the prize jointly with Christiaan Eijkman.

Szent-Gyorgyi

Hopkins and Funk's "unknown water-soluble substance" now had a name, vitamin C, and an assumed function, preventing scurvy, but its chemical composition was unknown. This began to change in 1928, when Albert Szent-Gyorgyi, a Hungarian biochemist who came to work in Hopkins's laboratory in Cambridge, isolated an acid from the cortex of the adrenal gland and found it to be a strong reducing agent. In his early years, Szent-Gyorgyi was an intuitive scientist of genius level who earned the nickname of "The Saint" amongst his co-workers.[35] He discovered that his new reducing agent, which he crystallised into a white powder, was related to sugars and was present in oranges and cabbage. When he submitted his results to the Biochemical Journal, he decided to call it ignose on the grounds that he was ignorant of its structure, and sugars are commonly given names ending in –ose: hence, ign-ose. The editor objected to this flippant name and asked Szent-Gyorgyi to change it. He did, but changed the name to "godnose". The editor did not share Szent-Gyorgyi's sense of humour and decided it should be called hexuronic acid instead.

Sixteen years after Hopkins and Funk proposed their vitamin hypothesis, Szent-Gyorgyi isolated hexuronic acid. He did not show that hexuronic acid was vitamin C immediately, although in his Nobel Prize lecture he states that he had suspected that the substances were the same "from the beginning". He returned to his native Hungary where he recruited a young American researcher. Svirbely "had experience in vitamin research, but besides this experience brought only the conviction that my hexuronic acid was not identical with vitamin C." Together, Szent-Gyorgyi and Svirbely carried out tests to see if hexuronic acid was anti-scorbutic, and, when they found that it was, they continued with the work in order to replicate the results. On obtaining a positive result, Svirbely communicated his findings to Charles King of the University of Pittsburgh. This was unfortunate, because in 1932, King and a co-worker, W.A. Waugh, published a paper in the journal Science suggesting that Szent-Gyorgyi's hexuronic acid was actually vitamin C. The King paper appeared 16 days before the paper in Nature by Svirbely and Szent-

Gyorgyi, who was not pleased that King and Waugh had beaten his paper to publication.

Despite losing the race to publish, Albert Szent-Gyorgyi was awarded the 1937 Nobel Prize in Medicine, for his discoveries in connection with biological combustion processes, with special reference to vitamin C. On this, the American press suggested that he had stolen the discovery from King and Waugh and that the prize should have been awarded jointly. However, it turned out that no one had nominated King and Waugh, and the Nobel Prize committee do not put forward candidates for the prize themselves. The committee was therefore unable to make a joint award.

Vitamin C may have been isolated and identified much earlier. As is often the case in science, this discovery was preceded by similar observations. Dr Andreas Mollenbrok probably isolated a crude preparation of vitamin C in the middle of the 17th century. Mollenbrok extracted a salt from a member of the cabbage family, called scurvy grass. Sailors ate the sharp-tasting leaves of this plant, which are high in vitamin C, to prevent scurvy. Scurvy-grass ale was a popular tonic drink. Mollenbrok claimed his salt, which he called the volatile salt of scurvy grass, was the active ingredient.[20] He also noted that descriptions of this salt could be found in earlier literature, on the use of this plant as a treatment for scurvy.

Szent-Gyorgyi went on to discover part of the citric acid cycle, a core element in the biochemistry of metabolism. Hans Krebs used Szent-Gyorgyi's results to complete the cycle that now bears Krebs' name and, again, Szent-Gyorgyi felt disheartened. He then turned to the biochemistry of muscle and laid some of the foundations necessary for the discovery of the mechanism of muscle contraction. He subsequently became involved in politics and called for Hungary's withdrawal from the Second World War. His most notable political achievement was to get Adolf Hitler personally to demand his head.

After the war, the quality of Szent-Gyorgyi's scientific output dropped dramatically. His arguments became confused and his ideas on the role of free radicals in biology and disease have been largely discounted. His isolated position, as the head of the Woods Hole Research Institute, was somewhat separated from the demands of normal science. Perhaps it was just old age but, while he clearly had a strong continuing belief in the biological role of free radicals and antioxidants, the standard of his work was poor.

Throughout his career, Albert Szent-Gyorgyi believed in the efficacy of vitamin C for promoting health. According to him, a medical conspiracy against vitamin C dates back, at least, to the isolation of the vitamin.[35] This viewpoint could be considered unnecessarily adversarial. An alternative explanation is that there is a basic difference in philosophy between conventional clinicians and those promoting alternative therapies. Medicine is inherently conservative: what appears like misleading the public, to an outsider, may be perceived as necessary prudence by clinical authorities.

The controversy begins

Albert Szent-Gyorgyi isolated vitamin C in 1928 and received his Nobel Prize in 1937. Around the same time, in 1932, Irwin Stone began a lifelong study of vitamin C, which he referred to as ascorbate. He worked as a chemist for the Wallerstein Company in New York and was initially interested in the anti-oxidant properties of the newly discovered vitamin, as a means of protecting food against deterioration. By the 1950s, Stone had become convinced that humans could benefit from far larger doses of ascorbate than were needed to prevent scurvy.

In April 1966, Irwin Stone met Linus Pauling, who found him to be well informed and convincing. Pauling credits this meeting as the starting point for his own interest in vitamin C. After looking at the evidence, Pauling was convinced that there was a case for vitamin C supplementation. He wrote his book on vitamin C and colds, perhaps naïvely expecting that the medical profession would be pleased to get the information. If it were true that vitamin C could prevent, cure, or even just ameliorate the common cold, he thought, much suffering might be avoided and physicians could get on with curing serious diseases. The result, although astounding, was far from what he expected. The medial profession vilified Pauling, branding him a quack and a charlatan. It was claimed that he did not know what he was talking about, with the implication that a mere chemist could not understand the intricacies of medicine.

On the other hand, the public were in awe of Pauling and his achievements. Their implicit reasoning can be stated as follows: if Linus Pauling, the world's greatest chemist, thinks that vitamin C is "the most important substance in the world" then there must be something in it. Of course, the opinion of someone with an outstanding reputation can be mistaken and the history of science provides numerous examples of gaffes made by great men. To put this into context, however, Linus Pauling was one of the greatest scientists who have ever lived. He was probably the leading American scientist of the 20th century and certainly one of the most well known. He gained two unshared Nobel Prizes, the first for chemistry, and the second for peace.

Apart from having revolutionised chemistry, at the time he wrote his book on vitamin C, Pauling had been publishing groundbreaking biomedical research for over thirty years, since 1934. Indeed, he received 37 medical awards, some of which would be the pinnacle of achievement for a world-leading physician. His findings included showing that sickle-cell anaemia is a disease of haemoglobin, the first known molecular disease. Pauling also predicted that enzymes would work on a "lock and key" mechanism and this was confirmed many years later. He discovered the basis of three-dimensional protein structure: the hydrogen bond, the alpha helix and pleated sheets. He also published a molecular theory of anaesthesia. These biomedical achievements alone would make him a leading medical scientist but they were arguably minor compared with his work in chemistry. Despite these monumental achievements, Linus Pauling would have been the first to tell you that, when considering vitamin C, the only important thing is the evidence.

From a scientific standpoint, it appears surprising that the medical establishment attacked Pauling and his ideas so strongly. One might wonder at these physicians' scientific backgrounds, given that they were not embarrassed to level their accusations at a scientist of Pauling's standing. In his book, Pauling presented a hypothesis with a basis of experimental support. Why was he attacked? We can only guess: perhaps the medics were upset that the book sold so well and caught the public imagination. Maybe they were jealous. Perhaps the drug companies were worried that people would use fewer cold medications. Who knows? Pauling was, however, a capable opponent. After decades at the peak of science, he could defend his corner. In the face of this challenge, the medical establishment closed ranks and since then, vitamin C research has been viewed as objectionable and largely unworthy of serious consideration. Some have even argued that Linus Pauling went through the same process as Albert Szent-Gyorgyi, claiming that a long career of outstanding achievement was flawed as Pauling became older and "lost the plot" with vitamin C. There is little evidence for this view.

One unfortunate legacy of the Pauling - vitamin C controversy is that it has entrenched opinion so strongly. The medical profession depend on authorities for information about current best practice. These medical authorities are understandably cautious when it comes to innovation and new ideas. General practitioners, for example, must struggle to keep track of the mountains of information on new drugs and treatments. Guided by consensus and current medical opinion, if the accepted policy is that vitamin C supplements are useless, then that is what general practitioners are likely to believe. From his position as the world's leading chemist, Pauling could call on no authority other than his own intellect. To understand why he staked his reputation on vitamin C, we need to know a little chemistry.

Chemistry and vitamin C

**"Chance favours the prepared mind."
Louis Pasteur**

Vitamin C is an antioxidant. Such substances are widely promoted as beneficial, in products ranging from face cream to pet food. To appreciate the function of antioxidants, we have to know some chemistry. This simplified account provides the background for understanding the health benefits of ascorbate.

Oxygen

We might guess from the name that *antioxidants* have something to do with oxygen, the gas that makes up about a fifth of the earth's atmosphere. It is common knowledge that oxygen is essential to life, but less well known that it can also be a deadly poison. To understand this paradox, we must look at the chemistry of oxygen as it relates to living beings such as ourselves.

Plants produce oxygen as a by-product of photosynthesis, the process by which they use energy from sunlight to build glucose from carbon dioxide and water. This process ultimately provides the energy for most life on earth. Both plant and animal cells require oxygen to burn food and produce energy. A few life forms do not depend on oxygen to live: these often require a different source of free radicals, such as sulphur. The main exception is a group of anaerobic or oxygen-hating bacteria that cannot live in the presence of oxygen. These anaerobic bacteria do not have sufficient antioxidant defences to exist in an oxygen rich environment.

A fire needs oxygen to burn, producing heat energy along with products such as ash and carbon dioxide. Human life itself also requires energy, which our bodies produce by breathing in oxygen and metabolising food, in a process that we can describe as slow combustion. We harness the energy produced by burning food to maintain our body temperature or to power the reactions needed for muscles to contract. However, our own body tissues are not so different from the food we eat, and can be damaged by the same oxygen molecule that is so essential to us. In response to the damaging aspects of oxygen, our bodies have evolved a form of protection, based on antioxidants. Living organisms have a large array of antioxidant defence mechanisms to protect them from oxidative damage. Without these, life could probably not exist in our atmosphere.

Oxygen damages biological molecules by stealing electrons, a process known as *oxidation*. Pure oxygen injures and kills cells and tissues. Most diseases and features of aging involve damage to tissues caused by oxidation reactions, involving free radicals. Examples of oxidation include the process that makes sliced apples turn brown when left in contact with the air, or the explosive reaction of metallic sodium with water. Antioxidants, including vitamin C, are substances or processes that can prevent oxidation damage by donating electrons to replace the lost ones. This process is the reverse of oxidation and is called *reduction*.

The abundance of oxygen in the atmosphere comes from plant photosynthesis, and its presence means that oxidation and antioxidant defences have played a large part in the evolution of life on earth. Some biologists suggest that multicellular organisms originated because groups of cells clumped together to lower their oxygen load. Higher up the evolutionary chain, animals breathe oxygen in order to live. The oxygen we breathe allows us to burn our food, releasing its energy in a controlled set of biochemical reactions. At various stages in this process, energy is released in the form of electrons, which are used to power the cell. The energy processes in cells depend on transfers of electrons and protons, in chains of oxidation and reduction reactions.

It is not generally known how important protection from oxidation is to life. Later, we will see that oxidation is a consistent feature of disease and illness, as might be expected of a process that is so central to biology. The basic building blocks of our bodies are cells, consisting largely of water, protein, fat, and genes made of DNA (deoxyribonucleic acid). However, in order to exist and evolve, cells also need antioxidants. Biological cells are packed full of antioxidants, to protect themselves from a hostile, oxidising environment. The central position of ascorbate as an antioxidant is demonstrated by its ubiquity throughout the plant and animal kingdoms.

Molecular structure of oxygen

Molecules are composed of atoms, which consist of a nucleus of neutrons and protons, surrounded by a cloud of negatively charged electrons. As an electron spins, its electric charge generates a magnetic field. Most electrons in molecules exist in stable pairs. With two paired electrons spinning in opposite directions, the magnetic fields cancel each other out. Oxygen is a strange molecule however, as it is stable despite having two unpaired electrons. Because it contains two unpaired electrons spinning in the same direction, oxygen is magnetic. Magnetic molecules and atoms, such as iron, have at least one unpaired electron. The unique chemistry of oxygen derives largely from it having two unpaired electrons.

Oxidation and reduction

Despite the name, the process of oxidation does not require oxygen. Most people have a fair understanding of energy release from fire and oxidation. Living creatures also obey the laws of physics and chemistry. In metabolism, our food is oxidised in a controlled way, extracting the energy bit by bit, as we break down the complex food molecules. Our bodies take great care to keep the oxygen bound safely to molecules such as haemoglobin, which transports it in the blood. Too much free oxygen would oxidise our tissues and kill the cells.

Oxidation = losing an electron
Reduction = gaining an electron

A well-known example of reduction is the preservation of Lindow man and other bog-people, over thousands of years. Because a peat bog slows down the rate of oxidative damage, tissues may survive for extended periods with minimal degradation. Another example is the sacrificial anode: a block of metal, such as magnesium, which is used as an antioxidant to stop a ship's hull from rusting. An iron or steel hull will rust in seawater but if a different metal that oxidises more easily is attached, it can donate electrons to the hull. Since steel is a good conductor of electricity, the electrons can flow from the sacrificial anode to the hull, preventing rust. As the steel loses an electron in an oxidation reaction with the seawater, it takes a replacement electron from the sacrificial anode, which is oxidised instead of the hull.

Free radicals

A *free radical* is a molecule with one or more unpaired electrons. It will take electrons from any molecule it happens to bump into. Atoms and molecules in the body are constantly vibrating and moving, which ensures rapid chemical reactions. Free radical reactions are biologically important, as they cause tissue damage.

Free radicals can be oxidising or reducing agents. The more dangerous oxidising free radicals react with normal molecules and set up chain reactions. Reducing agents donate electrons to reduce other molecules, which can stop free radical chain reactions. Two free radicals can also react, forming a covalent bond and combining to form a single molecule. In thinking about disease, we are mostly concerned with oxidising free radicals.

These oxidising free radicals can be highly damaging to the body, producing a chain reaction as one molecule after another is oxidised. An antioxidant like vitamin C can donate an electron to a free radical, stopping the reaction by allowing the free radical's reactive electron to form a balanced pair. This protects cells by preventing free radical damage to essential molecules. Antioxidants are sometimes called *free radical scavengers,* as they neutralise free radicals in the body.

For simplicity, we use the terms oxidants, reactive oxygen species and free radicals quite loosely and interchangeably, as they can all lead to tissue damage by oxidation. There are precise scientific definitions, such as, "a free radical is a reactive part of a molecule or a chemical species with one or more unpaired electrons",[36] but our description is sufficient to appreciate the properties of vitamin C. The term *reactive oxygen species* covers a range of free radicals and molecules derived from oxygen. Non-radical reactive oxygen species include hydrogen peroxide, hypochlorous acid, ozone and singlet oxygen. Reactive species can also be based on other elements: reactive nitrogen and chlorinating species are found. This form of oxidation is common in normal life, for example in antiseptic cleaning and common household bleach.

X-rays and pure oxygen damage our tissues in similar ways. Pure oxygen, especially at high pressure, produces a massive increase in free radical reactions in sensitive tissues such as the brain. X-rays and gamma rays generally travel straight through our tissues, but occasionally they may hit our atoms and be absorbed. Without this absorption, an x-ray image would not form. If we ignore absorption by bone, water molecules absorb most x-rays, since our bodies are largely composed of water. X-rays can damage protein, fat and DNA molecules by hitting them directly, but more damage is caused by the free radical chain reactions that originate with water.

Both x-rays and oxygen can split water molecules, producing free radicals. The damaging free radicals that are of central importance to biology are, ironically, intermediates between two molecules that are essential to life: water and oxygen. At first, it might seem strange that two things as apparently different as oxygen and x-rays harm the body in essentially the same way. Later, we will find that ageing, disease, stress, chemicals, and other insults to the body also ultimately cause damage by means of oxidation and free radicals.

Ionisation is the process by which a molecule loses or gains an electron to form a free radical. X-rays are a form of *ionising radiation*, as they knock electrons out of molecules. In the nuclear centre of atoms, protons have a positive electrical charge, while neutrons have no charge. A cloud of electrons, each with a negative electrical charge, surrounds the nucleus. In normal circumstances, the electrical charge on the nucleus balances and is shielded by the charge on the electrons. The forces holding the nucleus together are much stronger than are those holding the electrons. Nuclear power depends on the high energies in the nucleus of the atom, whereas chemical and biological reactions involve the lower energy interactions of the electrons.

Reactive oxygen and nitrogen species

Without oxygen, people die rapidly, but this element is also highly poisonous. Even the low concentration of oxygen in the atmosphere, about 21%, is damaging to air breathing creatures. The oxygen you are breathing as you read this sentence is damaging your tissues. Higher concentrations of oxygen than those found in air are generally toxic to both plants and animals. Even simple bacteria will swim away from areas of high oxygen concentration to achieve a more acceptable *redox* state. The redox state is the complementary balance between **red**uction and **ox**idation.

The damage caused by high oxygen concentrations is of interest in medicine because of the concentrations used in hyperbaric oxygen chambers. It is also of importance in other areas such as deep sea diving, submarines and spacecraft. In the early 1940's, a new disease called retinopathy of prematurity was described, which involved formation of fibrous tissue behind the lens in the eye. In 1954, researchers realised that this disease was a result of the high oxygen concentrations used in incubators for premature babies. Nowadays, doctors monitor the concentration of oxygen in incubators and antioxidants may be used.

Another example of damage caused by oxygen is in occlusive stroke and coronary occlusion. In stroke and heart attack, arteries become blocked by blood clots and tissues supplied by the blood vessel die when deprived of oxygen. Surprisingly, damage also occurs when the blood supply is restored. The re-supply of oxygen triggers a massive burst of free radicals that kills and damages cells. Some antioxidants, such as lipoic acid and vitamin C, can protect against this reperfusion damage.[37]

The hydroxyl radical

Water covers two thirds of the surface of the planet and its properties are well known. The water molecule, H_2O, consists of two atoms of hydrogen and one of oxygen. Water molecules can split back into their constituent parts in different ways. Conversely, when oxygen combines with hydrogen to form water, the same free radical intermediates can occur. These are biologically important.

For example, when an x-ray knocks an electron out of a water molecule, the molecule can split into a hydrogen ion, H+, and a hydroxyl radical, ·OH. The hydrogen ion exists normally in solution and merely adds infinitesimally to the acidity of the water. By comparison, the hydroxyl ion (·OH) is an extreme free radical: it will try to get an electron from anywhere it can. The dot (·) on the OH of the hydroxyl radical indicates an energetic unpaired electron. These molecules react instantaneously with any biological molecule they meet on their rapid diffusion through our tissues. Hydroxyl radicals are so reactive that dietary antioxidants are usually considered ineffective in stopping their destructive action.[36] The antioxidant would need to be in the tissues at a high concentration, to have any chance of being the first important molecule the free radical meets.

In the absence of massive doses of antioxidants, a hydroxyl radical can trigger a sequence of damage. If the hydroxyl ion hits a protein, it steals an electron and combines with a hydrogen ion to become an inert molecule of water once again. However, the protein is now one electron short and has become a free radical in turn. The protein wants to replace its lost electron and will steal one if it can. This is how a free radical chain reaction starts. A free radical steals an electron and becomes stable. The molecule that has had its electron stolen is now a free radical and will steal one in turn. The process continues, until it reaches a molecule that gets an electron by reacting with itself or by combining with another molecule to produce a corrupt molecular structure, that has lost its biological function. Butter becomes rancid in this way. In rancid butter, the fats have changed their form in response to free radical attack. This oxidation of fats is called peroxidation and, as we shall see, it is a central feature of heart disease and strokes.

From oxygen to water

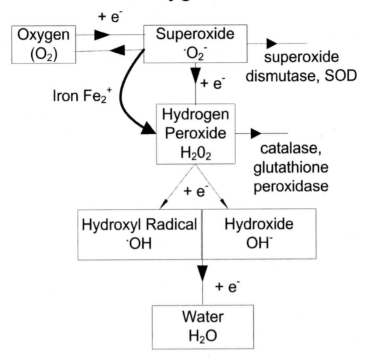

Hydrogen peroxide

If water loses two electrons, it can form *hydrogen peroxide*, H_2O_2. This chemical is commonly used as a bleaching or oxidising agent and the name has given us the "peroxide blondes" of Hollywood. In a sense, hydrogen peroxide is positioned two electrons away from both water and oxygen. For this reason, it can act as either a reducing agent or an oxidising agent. It can gain electrons and be reduced back towards water, or can lose electrons and be oxidised towards oxygen. It can even react with itself to produce oxygen and water.

Normally, hydrogen peroxide acts as an oxidising agent. It becomes especially dangerous in the presence of iron. The reaction is sufficiently important that it has been given the name the Fenton reaction, after the chemist Henry Fenton who described it at the end of the 19th century. The problem with the reaction of hydrogen peroxide with iron is that it produces destructive hydroxyl radicals.

Hydrogen peroxide is highly reactive and converts to oxygen and water in the presence of an enzyme called catalase. Catalase was one of the first enzymes to evolve, as early life forms required antioxidant protection from their environment. We will have more to say about this enzyme when we discuss the use of vitamin C in cancer treatment. Another enzyme, requiring the antioxidant glutathione as a co-factor, can also destroy hydrogen peroxide. Vitamin C is not the most important antioxidant in the cell: that distinction goes to glutathione. Vitamin C and glutathione exchange electrons in a cellular network of antioxidant defences. However, since glutathione is a peptide, or small protein, it is broken down in the gut and is not suitable as a dietary supplement.

Superoxide

If we take an additional electron from hydrogen peroxide (and lose the hydrogens), we get the final intermediate between water and oxygen. This intermediate is called the superoxide radical, $\cdot O_2^-$. Despite its powerful sounding name, it is not very reactive, unlike hydrogen peroxide. Superoxide reacts rapidly with other free radicals but reactions with fat, protein or DNA are limited to acid conditions. In the presence of iron, superoxide can participate in a Fenton reaction to produce the vicious hydroxyl radical.

An enzyme called superoxide dismutase, or SOD in its poetic shortened form, converts superoxide back to oxygen and hydrogen peroxide. In nature, there are several different forms of SOD. Oral forms of SOD can be found in health food stores, but are probably destroyed in the stomach before entering the body. Enzymes like SOD are protein molecules, which the digestive system consumes and destroys; there are certainly more effective antioxidant tablets.

Structure of Vitamin C

Vitamin C is a small molecule, similar in structure to the sugar glucose. It is composed of six carbon atoms, six oxygen atoms and eight hydrogen atoms, all linked together by chemical bonds. It is a weak acid, also known as ascorbate and l-ascorbic acid. Many food supplements use the salt forms: sodium ascorbate, calcium ascorbate or magnesium ascorbate; these are neutral or slightly alkaline rather than acid and are easier on the stomach. Solutions of ascorbate are stable at room temperature, unless transition metals such as copper are present. Copper will oxidise ascorbate. So, if you want vitamin C from your food, do not cook your vegetables in copper pans.

Vitamin C as an antioxidant

Ascorbate has many actions in the body, but one of the most important properties is that it is an antioxidant. When a dangerous free radical reacts with vitamin C, the free radical gains an electron and becomes unreactive. The vitamin C itself is oxidised, since it has lost an electron. Ascorbate donates an electron to the free radical and forms semidehydroascorbate, also known as the ascorbyl radical.[38] Fortunately, the ascorbyl radical is unreactive and is neither strongly reducing nor oxidising. This low reactivity is the key to many of vitamin C's properties. Since the ascorbyl radical is unreactive, the free radical reaction is stopped or quenched.

Two ascorbyl radicals can combine, forming one molecule of ascorbate and one of dehydroascorbate.[39] Dehydroascorbate is ascorbate that has lost two electrons – it has been doubly oxidised. In many reactions, ascorbate donates two electrons to become dehydroascorbate, which can be reduced back again to ascorbate. The oxidised form, dehydroascorbate, is not stable and breaks down in a complicated way, forming oxalic and l-threonic acids.

ascorbate ⇌ dehydroascorbate

Ascorbate has a wide range of antioxidant properties outside the body and can quench most biologically active free radicals.[40] It scavenges superoxide, nitroxide, hydroxide and hydrogen peroxide and will reduce vitamin E. The list of properties of ascorbate as a reducing agent in the laboratory is long and is not reproduced here, as the chemical properties of ascorbate are not disputed. Inside the body, ascorbate has a large number of antioxidant roles but it is harder to demonstrate this directly.

Vitamin C as an oxidant

Many antioxidants normally present in our cells are reversible: they can both donate and accept electrons. When they donate electrons, they act as antioxidants. When they accept electrons, they act as oxidants and are themselves reduced. Vitamin C can act as an oxidant, especially in the presence of metal ions such as iron.[41] Many other biological antioxidants can also become oxidants, under the right conditions. Under certain circumstances, vitamin C can act as a pro-oxidant and damage cells. This is not always a bad thing - for example, vitamin C might oxidise cancer cells, killing them and leaving normal cells healthy. As we shall see, scientists claim that massive doses of vitamin C kill cancer cells, apparently by producing hydrogen peroxide. In this case, oxidation is clearly beneficial.

In the body fluids, ions of metals, such as copper and iron, are normally bound to proteins. However, this might not always be true inside the living cell. Some patients with iron overload diseases can have free iron in the plasma, and rare reactions to vitamin C administration have been reported in such cases. The idea that ascorbate can act as an oxidant under some circumstances is a more valid argument against high dose administration than the more frequent suggestion that it might cause kidney stones, but the supporting evidence is insubstantial. Vitamin C normally acts as a reducing agent.

Why are free radicals important?

The free radical theory of aging states that free radicals are the fundamental cause of aging. Adding antioxidants to the diet of animals increases their average life span, although generally, it does not produce an increase in the maximum life span of the oldest individuals. If mice live to be a maximum of three years old, mice given antioxidants will die closer to this age, but typically will not live longer than three years. Scientists have explained this finding by suggesting that antioxidants reduce the level of disease and degenerative processes but do not affect the core aging process.

The free radical theory suggests that aging is an oxidative process and provides a fundamental mechanism, linking several alternative explanations. The mitochondrial theory is a special case of the free radical theory,[42] in which the mitochondria, being a constant source of superoxide and other free radicals, are particularly susceptible to damage over time. Mitochondria are sub-cellular particles that provide the cell with energy. They are thought to have originated early in evolution, when a cell engulfed a bacterium and both survived.

Many free radicals are produced in the mitochondria, which slowly become damaged with time. As the mitochondria are injured by the oxidants they produce, their function declines and they produce even more free radicals. According to this model, the cell dies either because the mitochondria cease to provide it with sufficient energy, or because they poison it with oxidants. Leakage of reactive molecules from mitochondria is usually small and is roughly balanced by natural free radical scavenging mechanisms. The transfer of electrons in mitochondria is not completely efficient.[43] Perhaps one to three percent of the oxygen we breathe is normally converted to superoxide. This is a relatively efficient process, but the released free radicals need to be quenched by antioxidants before they damage the cell. Most dietary antioxidants do not reach the mitochondria and are not able to prevent this free radical damage. However, the antioxidants r(+)-lipoic acid and acetyl-l-carnitine have been shown to restore mitochondrial function in aged rats back to levels normally found in young animals.[44,45,46] Bruce Ames, one of the researchers, described the response, saying it was as if the supplemented old rats got up to "dance the Macarena"!

Free radicals and oxidation are involved in most disease processes. Just about every disease mechanism seems to involve free radical reactions, as do the body's defence mechanisms. However, antioxidants help bias the body towards a reducing state. The abundance of active antioxidants in the reduced state limits the damage to the tissues and promotes good health. As we will show in the following chapter, the use of antioxidants to protect against illness started a very long time ago.

Evolution and deficiency

"The mystery of the beginning of all things is insoluble by us, and I for one must be content to remain an agnostic." Charles Darwin

Vitamin C is an important antioxidant throughout the plant and animal kingdoms. Because it is so widespread, one way of estimating human requirements is to find out how much is used by related species of animals. Irwin Stone compared the diets and vitamin C intake of animals to that of humans, and concluded that humans were deficient. The first thing to note in such comparative studies is that humans are unusual because, unlike most other animals, we are unable to make vitamin C in our bodies.

Mutation

About 40 million years ago, our ancestors lost the ability to manufacture vitamin C in their bodies.[47] The ancestors concerned were small, furry, mainly vegetarian mammals, more like tree shrews than apes. Estimates place the date of this mutation in a range from twenty to sixty million years ago. In evolutionary terms, this date is not much later than the extinction of the dinosaurs.

Biologically, ascorbate is synthesised from glucose in a series of steps catalysed by enzymes. Enzymes are chemical substances produced by living cells, which promote particular chemical reactions while not themselves being changed. In humans, the gene for gulonolactone oxidase, the fourth and final enzyme required to convert glucose to ascorbate, is defective. The coding for this enzyme is present in human DNA, but the gene is mutated. This alteration could have been a result of the action of a retrovirus, the class of viruses that includes the human immunodeficiency virus (HIV), believed to cause AIDS.[48]

There is little evidence concerning the evolutionary consequences of the loss of the ability to make the vitamin, although it is thought that such a loss leads ultimately to a reduced biological fitness. Since the descendants of the individuals in whom the mutation occurred did not die out but prospered, it has been assumed that they must have been consuming a diet high in vitamin C. Man has many other antioxidant defence mechanisms, but they do not protect us from the same free radicals as vitamin C. All descendants of the individuals carrying the defective gene are therefore dependent on getting ascorbate from their diet. For this reason, scurvy can be viewed as a genetic disease, an inborn error of metabolism.[47]

It is often stated that man, monkeys and guinea pigs are the only animals that are unable to make their own vitamin C, but this is a simplification based on inadequate data.[49,50,51] Most animals have not been tested for the ability to manufacture vitamin C. It is not even known for sure that all humans have completely lost this ability. The closer we examine the available data, the more complex the situation appears. We now have evidence that dietary vitamin C is required by bats, some fish, many birds but not all primates. Animals that make their own ascorbate do so in different organs. Mammals and perching birds make it in the liver, while reptiles and other birds tend to make it in the kidney.

Implications of gene mutation

Irwin Stone became interested in the amounts of ascorbate present in the tissues of animals other than man. Animals that synthesise ascorbate make relatively large amounts, although the evidence is incomplete. Rats are reported to make 70mg per kilogram of body weight each day.[52] If the rat is stressed, the amount of ascorbate it produces increases by a factor of about three, to 215mg per day.[53] Plasma levels of ascorbate reduce to one half (from 56 to 29 µM), and excretion increases by a factor of 10 (from 46 to 450 µM) in diseased rats.[54] Injection of ascorbate (equivalent to about five grams in a human) restores plasma levels, blood pressure and capillary perfusion to normal levels; ascorbate also inhibits growth of bacteria. Rats respond to infection by increasing both synthesis and excretion of ascorbate. Similar stress related increases are found in other animals that produce their own vitamin C.[55]

Weight for weight, the rate of ascorbic acid production in the rat is equivalent to between 5 and 15 grams per day given intravenously to a 70kg adult human. By comparison, 10 grams is thought to be a very large dose of vitamin C for a person; the recommended daily allowance (RDA) is only 60mg per day. It is possible that the requirements for rats are different to those for man. However, similar rates of production are found in goats (human equivalent 13 grams) and other animals. Domesticated cats and dogs produce less (human equivalent 2.5 grams). It can be misleading to extrapolate from animals to man. There are some important differences between closely related animals, despite the fact that unrelated animals, such as worms and elephants, have remarkably similar biochemistry. Even within a species, the biochemical variation between individual animals can be large.

Where animals, such as guinea pigs and bats, have lost the ability to make ascorbate, it has been claimed they eat a diet that provides a high level of the vitamin. Once again, the evidence does not support such a strong statement, as we do not have a complete list of these animals. However, in the case of our nearest relatives, the primates, we know that they consume large amounts of vitamin C in their mainly vegetarian diet. Primates may not have a diet that provides as much vitamin C as animals that manufacture the substance internally, but it is clearly enough for evolutionary success. A laboratory monkey requires the equivalent of one gram of vitamin C per day, whereas a human is claimed to need only a fraction of this amount.[2] A modern gorilla, living naturally in the wild, eats about 4.5 grams of ascorbic acid per day.[2] Linus Pauling used the nutritional values of plant foods to estimate that, assuming an early human diet similar to that of the great apes, the human intake of ascorbate should be 2.3 to 9.5 grams per day.[56]

Deficiency

Assuming our biochemistry is similar to that of our nearest animal relatives, we should be consuming large amounts of vitamin C each day and most people are therefore deficient. According to Pauling, the range of the B vitamins in 110 raw plant foods supplying 2,500 calories is two to four times the recommended dietary allowances. However, the corresponding amount for ascorbate is at least 35 times the RDA. A vegetarian eating raw plant food would ingest a much larger quantity of ascorbate than of the B vitamins.

Although this is an interesting argument, we have little information about the evolution of man and what data we do have derives largely from skeletal remains dating from the last two million years. The mutation happened about 40 million years ago and we have no information about the lifestyle of our distant ancestral species. While it is likely that our ancestors were largely vegetarian and the plants they ate had similar levels of vitamin C to those we find today, we cannot be certain about this.

Surprisingly, when our ancestors' ability to make vitamin C was lost, it might not have reduced their evolutionary fitness. If they were consuming high levels of vitamin C in their diet, they would not have needed to manufacture it internally. Indeed, as Pauling has pointed out, not making ascorbate at the time of the mutation may even have had survival value, as it modified biochemical pathways and conserved energy.[57] Robert Cathcart goes further, explaining that in times of food shortage, animals that do not convert scarce glucose into ascorbate have a clear advantage and out-starve those that do.[58] This increased survival value would have been lost when descendants of our pre-human ancestors changed their diet to include less vitamin C.

Pauling's and Cathcart's suggestions are plausible, but it is not necessary for a mutation to add to the survival value of an organism for it to continue in the population. It is possible for a mutation that is neutral or even slightly deleterious to exist in a population over a long period. The classic example of a harmful gene becoming established in the population is the gene for sickle cell anaemia. Linus Pauling discovered the cause of this disease in 1949. A gene coding for a protein in haemoglobin, the substance that carries oxygen round the bloodstream in red blood cells, is damaged. The sickle cell gene is damaged by a single mutation in which an amino acid (glutamic acid) is replaced by another (valine). A homozygous person, with two copies of the defective gene, suffers sickle cell anaemia, a serious blood disease. However, a heterozygous person, with only a single copy of the defective gene, has enhanced protection against malaria. In places where malaria is prevalent, the sickle cell gene may be present in a high proportion of the population. The sickle cell illness is balanced by the protection against malaria, in a biological cost benefit analysis.

Hypoascorbemia

Following on from his evolutionary studies, Irwin Stone suggested that present day humans suffer from an inborn error of carbohydrate metabolism called *hypoascorbemia*.[59,60] This means a chronic disease resulting from too little ascorbate in the diet. While scurvy is an acute, life-threatening illness, hypoascorbemia is the result of an ongoing deficit, in which there is enough vitamin C to prevent scurvy but not enough to prevent disease in the longer term.[61] According to Stone, ascorbate has been considered as dietary "vitamin C", whereas in most animals it is made internally in large quantities, and should not be regarded as a vitamin.

The counter argument to Stone's suggestion is that since the original mutation, humans could have adapted to low levels of ascorbate. The evidence to support this suggestion is that above an oral intake of about 200mg, ascorbate is excreted by the kidneys. A daily dose of only 40-60mg is enough to maintain the tissue levels found in the population and, in the UK, the reference intake for adults in 1991 was 40mg per day.[62] This intake prevents acute scurvy. Nonetheless, these amounts are well below the 100-200mg per day threshold at which "excess" ascorbate is excreted. The sceptical argument can be summarised as implying that if doses above about 200mg are excreted in urine, anything above this is a waste of money. This claim is incorrect. It assumes that animals that make their own ascorbate do not excrete it, which is not the case.

Pauling's evolutionary argument supports a megavitamin hypothesis but, although interesting, it is not strong evidence in scientific terms. During the 40 million years since the loss of the ability to make vitamin C, evolutionary pressure could have reduced man's dietary requirements to milligram levels. Indeed, Pauling's own therapy for heart disease, considered in detail later, depends upon a form of cholesterol having evolved to replace vitamin C in the healing of arteries. Clearly, none of these evolutionary approaches tells us how much vitamin C we need in our diet. They also provide little hard evidence concerning the amount required each day. We will investigate this question further in the next chapter.

How much does a healthy person need?

"The establishment defends itself by complicating everything to the point of incomprehensibility."
Fred Hoyle

The question "How much vitamin C does a healthy person need?" sounds relatively straightforward. Scientifically, it is a surprisingly difficult question to answer and, in reality, there is insufficient evidence to reach a reliable recommendation for daily intake.

Recommended dietary allowance (RDA)

The US Food and Nutrition Board have prepared Recommended Dietary Allowances since 1941. Initially, the RDA was based on the amount needed to prevent people getting scurvy, which can easily be measured in short-term studies. The dose required to prevent acute scurvy was established many years ago, as a few milligrams per day.

In the United States, the RDA is defined as "the level of intake of essential nutrients that, on the basis of scientific knowledge, are judged by the Food and Nutrition Board to be adequate to meet the known nutrient needs of practically all healthy persons". Individuals with special needs are excluded. It should be understood that RDA values are not scientific, despite claims to the contrary, and many professional scientists and clinicians question their utility and significance.

The RDA is said to be the amount of vitamin that provides the *least risk of inadequacy* and the *least risk of toxicity*. Unfortunately, this definition is too vague to have any scientific meaning. The phrase "least risk" is fine if it applies to an estimated probability. However, the work underlying the RDA does not have enough data to estimate probabilities accurately. The word "inadequacy" is undefined: it could mean not dying of acute scurvy or, on the other hand, it could imply a reduced risk of heart disease, cancer, cataracts or a thousand other problems in the longer term. The toxicity depends both on the dose and the biological variability. For any given person, the required amount is a result of a cost benefit analysis and, as we shall see, there is insufficient data to make a global recommendation based on scientific evidence.

Biological variation

In the definition of RDA, it is claimed that the recommendations are a result of "scientific knowledge". However, the philosophy behind the RDA concept is based on outdated biological ideas. Evidence from the last fifty years does not support the suggestion that a single vitamin C recommendation is suitable for a biologically variable population. Furthermore, the RDA values have been determined according to methods that tend to overlook individual variability. The inadequacy of such approaches was made clear as long ago as 1956, by Roger Williams in his classic book "Biochemical Individuality".[5] With the arrival of technologies such as genetic sequencing, has come the expectation that medical treatment and nutrition will soon be tailored for our individual biochemistry. Conventional RDA values persist because they provide a simple intake level that can be used to prevent acute deficiency.

Chronic deprivation

In addition to concerns about biological variability, the RDA does not differentiate between short and long-term effects of different levels of deprivation. For example, chronic disease might occur, even with a vitamin C intake several times greater than the RDA. In scientific terms, the suggestion that sub-clinical scurvy causes chronic disease is a valid hypothesis, although admittedly it might be difficult to investigate. Studies of chronic disease and ageing can take years to complete, whereas much of modern science depends on achieving quick results.

To follow the long-term effects of different levels of ascorbate, even on short-lived animals, would be expensive, as it would take years to complete the project. One possible research subject, the guinea pig, can live from four to seven years in captivity, whereas by most research and funding standards, five years is a long-term project. A longitudinal study of the effects of varying levels of vitamin C intake in humans over, say, a 70 year lifespan, would take several generations of scientists, even if the subjects were prepared to give permission and stick to the dietary protocol.

The general hypothesis that chronic disease arises from sub-clinical vitamin C deficiency is weak, because it makes no specific predictions and would be difficult to evaluate experimentally. Refuting the proposal for a particular disease, such as arthritis, would provide no evidence for the effects on heart disease, for example. Yet, if true, the idea that sub-clinical scurvy causes chronic disease has enormous implications for health.

RDA based on biochemistry

To get round the chronic disease question and address the problem of setting a more reliable RDA, one approach has been to measure the absorption and excretion of vitamin C. Following criticism of the scurvy prevention method for determining the RDA, Dr Mark Levine, from the US National Institutes of Health, proposed that requirements for vitamin C and other vitamins could be determined using biochemical measurements.[63,64,65] Levine hoped to put nutrition into a biological context, by determining a person's requirements from the amount needed for essential chemical reactions. In his initial experiments, the aim was to find an optimal intake by measuring how much is absorbed, retained in the tissues or excreted.

The reasoning behind these experiments goes something like this: if the person is consuming too much vitamin C, then any excess will be excreted. If they are not consuming enough to maintain physiological levels, then active pumps in the kidneys will reabsorb vitamin C and conserve it for future use. These ascorbate pumps help prevent scurvy when the diet is deficient. However, while prevention of acute deficiency provides an evolutionary advantage, it does not preclude chronic disease in the older animal. It is a general feature of biology that evolutionary adaptation is greater over short periods and in younger, reproductively active animals. Therefore, absorption and excretion measurements by themselves may add little to our understanding.

Absorption and excretion

Knowledge about the absorption and excretion of vitamin C is rather limited and does not take into account individual variation or state of health. Specifically, most results come from small numbers of healthy young adults, so we do not know how very young, old or sick people respond. In healthy, young adults, vitamin C is actively removed from the gut[66] and at low doses, say below 60 mg, almost all is absorbed.[67] The proportion (though not the absolute amount) absorbed in a healthy individual decreases with dose: up to 80-90% of a 180mg dose is absorbed,[68] this reduces to 75% at 1 gram, 50% at 1.5 grams, 26% at 6 grams and 16% at 12 grams.[62,69,68] So, with a single dose of 12 grams, the actual intake into a healthy body is about two grams and the rest is excreted. Once in the body, the ascorbate is distributed widely and some tissues, such as brain, white blood cells and the retina, contain high concentrations.

The way vitamin C is used and excreted by the body depends on the dose. In humans, the principal route for elimination of the breakdown products of ascorbic acid is excretion in urine. If the dose is low, only a small amount of the vitamin is excreted unchanged. For example, about 20-25% of a 60mg dose is excreted as ascorbic acid, some of which is in the oxidized form, dehydroascorbic acid; a further 60% is excreted in modified form as either diketogulonic acid (20%) or oxalate (40%).[70] With increased doses of ascorbate, correspondingly more is excreted unchanged. Approximately 60% of a daily dose of 180mg is excreted as ascorbic acid, 80% of a one gram dose and 90% of a three gram dose.[67,66]

Levine's work on recommended daily allowances

Mark Levine proposed recommending a dietary vitamin C allowance for healthy young men.[63] In a subsequent paper, he repeated the research with 15 healthy young females and obtained similar results.[65] His study measured the absorption, blood levels and excretion of vitamin C. This work has been accorded great significance and is cited as underlying the standard RDA for vitamin C.[63,65,71,72] The Linus Pauling Institute (but not Pauling himself) have also based their recommendations on Levine's paper.

We have examined these papers in some detail, as they represent establishment thinking in this area. For example, according to the National Academy of Sciences *Dietary Reference Intakes for Vitamin C, Vitamin E, Selenium, and Carotenoids* (2000): "The rigorous criteria for achieving steady-state plasma concentration (five daily samples that varied less than or equal to 10 percent) make the Levine data unique among depletion-repletion studies." The ready acceptance of Levine's work contrasts markedly with the rejection of Linus Pauling's work on large doses.

Mark Levine is a dynamic and outspoken researcher, with an open-minded approach to vitamin C. We will consider his first paper, which proposed that the RDA for men should be increased from 60mg to 200mg daily, in detail. The subsequent paper, which related to women, is similar and provides less detailed data. As we will show, this research does not relate to optimal dosing. Levine has made an honest attempt in a difficult and contentious area, but his analysis is necessarily limited by the need to conform to the aims and philosophy of the RDA.

Levine's paper recommends 200mg per day as the RDA for vitamin C, and adds that doses above 400mg provide no benefit. In our analysis, we will see whether these assertions are supported by the data.

The research provides basic data on how the healthy body handles vitamin C. Levine measured absorption of orally administered vitamin C and compared it with similar amounts injected directly into the blood. He found that higher doses of vitamin C were incompletely absorbed by the body. He also investigated the uptake of ascorbate by white blood cells. Specific details of this work are discussed below.

Number of subjects

Levine's first paper derived its recommendation for a recommended dietary amount from a study of just seven healthy, young, male volunteers, aged 20-26, who were in a hospital for a period of between four to six months. His second paper studied a similar set of 15 healthy, young female subjects. If Levine's RDA recommendations are to be taken at face value, the few subjects he studied have to represent the biological variation of the whole US population, about 280 million people, in response to vitamin C.

There is little doubt that Levine is aware of the difficulties in extrapolating from such small samples to the population as a whole. These are difficult, long-term studies and he was limited to a small sample by the intensive nature of his investigations. In Levine's experiments, the number of subjects is relevant, as the results are being applied to the whole population. Sampling a population generally produces a reduced variability, since extreme values are unlikely to be represented in the selected data.

Blood plasma levels

Levine measured what he called "steady state" blood levels of vitamin C. We will explain the meaning of this term. If a drug is administered repeatedly, at intervals that are short relative to the rate of removal from the body, it will accumulate. With time it will reach a plateau level or steady state. A standard method is used in pharmacology to obtain steady state levels of drugs in the blood stream. Heart patients, for example, may require a constant level of a beta-blocker for a consistent effect, without danger of too low or too high a dose. For long-acting drugs, the dose can be adjusted to build up to a steady state over a period of time.

To obtain steady blood levels, the dose interval has to be similar to the length of time it takes to excrete the drug. Consider a drug dose of 100mg, with a half-life of one day. This means that half the drug (50mg) will be excreted in the course of one day. If a further 100mg dose is taken on the second day, the effective dose is 100mg plus the 50mg that is still in the body from day one: a total of 150mg. If a third dose is given on day three, the effective dose is 100mg plus 50mg (from the previous day) + 25mg (from the day before): a total of 175mg. Over time, the blood levels reach a steady state. However, if the drug has a short half-life, say just one hour, then essentially all of the previous day's drug is removed before the next dose is given, and there is little build-up in the bloodstream.

Levine has tried to determine a steady state blood level for vitamin C, but has misapplied the technique by using dose intervals that were long compared to the half-life. In applying this technique, Levine appears to have forgotten that the half-life of vitamin C is very short. When we discussed this point with him, he explained his difficulty in getting the subjects to conform to a protocol that involved repeated multiple doses at short intervals. He was also constrained in the use of slow release formulations because of the confounding effects of the filler and packing on the availability of the vitamin.

The advice he received from the pharmacologist on the study was that a true steady state method could not be used with vitamin C, as a residual level was retained in the body over a long period of time. This advice was almost certainly incorrect, leading to a study design that was compromised. When we discussed our alternative interpretation with Mark Levine, he was remarkably helpful and suggested that he would recheck with his pharmacologist about the applicability of the techniques used. He added that he was considering the effects of closely repeated doses.

Levine determined his "steady-state" plasma and tissue concentrations at seven vitamin C dose levels, ranging from 30 to 2500 milligrams per day. Although the study covered a range of doses of vitamin C, it excluded the higher doses recommended by Pauling and others. The plasma measurement was made each day before subjects received their dose of the vitamin. This means the measurements were taken when the vitamin C had been excreted. These readings would measure the background level of vitamin C in the blood, as opposed to peak or even average values.

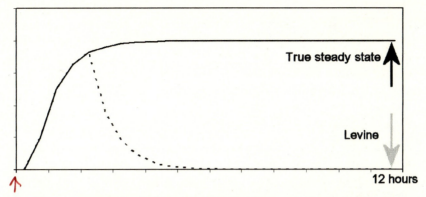

Plasma ascorbate levels, single dose vs. hourly doses

This schematic graph was computed from published vitamin C absorption and excretion data.[74] It shows the levels of ascorbate above base level. The dotted line is for a single two-gram dose taken at time zero and the grey arrow shows what Levine measured. The solid line is for a two-gram dose followed by repeated dosing (one gram per hour) and shows that a steady state is reached within a few hours. Notice how the steady state value is higher than the *maximum* value for a single dose. The darker arrow shows a true steady state measurement. This diagram illustrates Levine's unfortunate "steady state" error clearly.

The problem with this measurement can be stated simply. Levine gave a dose of vitamin C, waited until it had been excreted, and then measured the blood levels. The fact that this was done repeatedly is irrelevant. When given intravenously, vitamin C has a half-life of about half an hour in blood plasma. This means that the blood levels will be halved every 30 minutes. The samples in this experiment were taken about twelve hours from the previous dose. Twelve hours is 24 half-lives and the blood levels would in theory be reduced to $1/2^{24} = 1/16,777,216$. This is such a small amount that repeated doses at this interval would not be expected to accumulate.

Oral doses take longer to be absorbed, and consequently more time to be excreted. Levine's results indicate that, with the oral doses used, some limited accumulation did occur. Even so, the steady-state level at an interval of 12 hours would not deviate much from the background, regardless of the size of the dose. Any accumulation that did occur could be related to changes in levels of enzymes, or to how the vitamin was distributed through body tissues, but its significance is not clear. Levine's steady state values are merely the background or baseline plasma levels with repeated dosing, when the supplement has been excreted. At some point, the increase in Levine's steady state levels with dose declined, at which point he described the blood as "saturated".

The term *"saturation"*, as used by Levine, is misleading. In chemistry, the word saturation is used to mean a solution that contains all of a substance that is capable of dissolving. In other words, it is a solution of a substance in equilibrium with an excess of undissolved substance. This commonly used scientific definition would suggest to the uninitiated that when Levine talks about plasma saturation, he means that the blood holds as much vitamin C as it is possible for it to hold. This is far from true. The base plasma levels tended to stabilise at a dose of around one gram per day, at which point Levine claimed the plasma was saturated. This might be taken to imply that higher doses of vitamin C did not raise the blood levels; however, Levine's own data show that they do, as can be seen from the graph below:

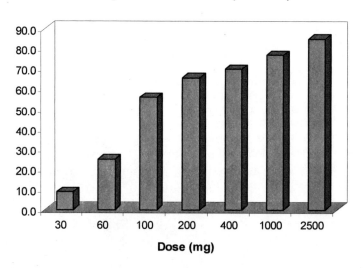

This graph (compiled from Levine's results in men) shows mean steady-state plasma values that increase over the range of doses. The highest plasma value is achieved with the highest dose of vitamin C. Furthermore, there is no reason to suppose that these residual plasma levels would not continue to increase with higher doses. It is evident that saturation has not occurred with a one-gram (1000mg) dose, since the value for a 2.5 gram dose is higher.

In a later paper, Levine essentially refutes his own idea that the blood plasma is saturated. He states that for a dose of three grams every four hours "pharmacokinetic modeling predicted peak plasma vitamin C concentrations of 220 micromol/L". In other words, repeated high doses provide a plasma steady state far in excess of his previous "saturation" measurements.[73]

What Levine calls the saturated value is actually the residual blood level for a repeated one-gram dose, after it has been excreted. The blood plasma is not saturated in the usual meaning of the word, as the levels would immediately increase above this level whenever a person took a further dose.

Use of the term "saturated" in this context is at best ambiguous, if not frankly misleading. A person taking divided doses or slow release tablets would consistently achieve plasma levels higher than the "saturated" value. The peak and average blood levels are biologically more important. These increase with vitamin C intake, especially for divided doses.[74] The use of what Levine calls "steady state" background levels has questionable significance, and is certainly not the solid evidence that those recommending the RDA have taken it to be.

White blood cell levels

Another strand of Levine's work was to measure the variation in what he called "tissue concentrations" of vitamin C, as a function of the dose level. The tissues in question were white blood cells: neutrophils, monocytes and lymphocytes. Levine found that, at low doses, white blood cells soon became saturated with vitamin C. The three white blood cell types measured became saturated with a dose of one hundred milligrams. It is important to point out that these data do not necessarily indicate that *all* tissues would be saturated at this dose. White blood cells accumulate and store high levels of vitamin C, even when levels in the surrounding plasma are low.[75,76] These cells have specific requirements for vitamin C in fighting infection. The difference between these white blood cells and most other cells in the body is illustrated by their short lifespan. From formation to death, white blood cells live for approximately one day and their death is controlled by antioxidants.[77,78] These cells are clearly specialized in their requirement for ascorbate.

Some cells in the body have biochemical pumps, which are able to concentrate vitamin C from the surrounding environment. There are at least three pumps for ascorbate and Mark Levine has done interesting work on their biochemistry. One pump transfers oxidized vitamin C into cells; this is used in phagocytic white blood cells, such as the neutrophils and monocytes used by Levine. Two others pump vitamin C into tissues that have a high requirement, such as epithelial tissues in the intestine, kidney and liver, or specialized cells in the brain, eye and other organs.[79,80] These pumps maintain high levels of vitamin C in specialized cells, which need it in times of deficiency. Although the pumps have a limited capacity for transporting the vitamin, the cells can become "saturated" at relatively low environmental ascorbate levels. At higher concentrations, the cells continue to accumulate ascorbate above this value but the rate is low and probably represents diffusion from the blood or plasma. The possession of ascorbate pumps provides an indication that white blood cells are particularly sensitive to depletion of the vitamin.

Neutrophils and monocytes are phagocytic, which means that they engulf foreign particles, destroying them with oxidants such as hydrogen peroxide. We will discuss the production and use of free radicals by white blood cells in more detail later, when we describe inflammation. Some lymphocyte white blood cells also contain the biochemical equipment for manufacturing free radicals. White blood cells concentrate vitamin C, because of their particular need for protection from oxidative damage. Biologically, this makes sense for an animal that is short of vitamin C. Concentrating and storing the vitamin in the white blood cells preserves essential immune function in the face of deficiency.

The preferential concentration of vitamin C in white blood cells is accepted by the Institute of Medicine as reflecting specific requirements.[64] Perversely, in justifying the RDA, the authors have then stated that these levels are a good estimator of total body levels. They do not provide evidence to support this statement. Even the study they use to sustain this argument states that these cells have particular needs. It relates the vitamin C levels in white blood cells to those in blood plasma, in a limited number of subjects at low doses, and indicates that further work is needed to estimate body status or the levels in normal tissues of the body.[81]

Mark Levine is not part of the RDA committee and has no responsibility for their actions. When we checked with him, he said that white blood cells were used in his study simply because they were easy to sample. He was not in a position to take biopsy samples from internal organs in a study of healthy people. He points out that other cells in the body contain pumps for vitamin C and white blood cells may represent some other cell types. Despite white blood cells having particular vitamin C requirements, they became a central part of the argument for the RDA.[64] Unfortunately, such tissues will be among the last to be depleted in times of shortage and are therefore least suitable for estimating general tissue requirements.

Since this is the only data presented by Levine for estimation of tissue requirements, we will take as a hypothesis that: "a proportion of normal, unstressed, young humans in their early 20's, who are in good health, can approach tissue saturation in some white blood cells with specialized requirements at a daily dose of only 100mg of vitamin C, if that dose is administered twice daily for an extended period." When stated carefully like this, we see the limitations of these results. We can accept that Levine's suggested 200mg RDA will provide steady state "saturation" of some specialized tissues that preferentially store the vitamin, in a proportion of healthy young people. However, his limited sample did not include old men, pregnant women, smokers, children, infants or the sick, for example.

The following anecdote illustrates how vitamin C requirements can vary, even within a single individual. Dr Riordan, a leading researcher into cancer and vitamin C, has described how the increased need in times of stress was one of the factors that led to his interest in the substance.[3] As part of a research study, Riordan's own vitamin C blood levels were being measured on a regular basis. His levels were in the typical range of 13 to 17mg per litre. However, during the time of the experiment, a spider bit him on the thigh; he was surprised to find that his blood levels subsequently dropped to undetectable values.

Thinking it would be easy to bring his levels back to the normal range, he asked a nurse to give him 15 grams intravenously: a large dose. The next day his blood levels were still undetectable. Amazed, Dr Riordan repeated the 15 gram injection, with the same result. He continued to repeat the cycle for five days, when his blood level became detectable but was still in the range typical of deficiency. It was several more days before his vitamin C level returned to normal and the spider bite was completely healed. Here, in one subject, we have an account of how a minor injury required more than 75 grams of vitamin C, given intravenously, before normal blood levels were restored.

This spider bite anecdote is reinforced by a French study of a patient in intensive care who was receiving 130mg supplements of vitamin C. Despite supplementation, this patient suffered scurvy when his ascorbate levels collapsed because of increased need following surgery.[82] It took four weeks of daily one gram doses to reduce the symptoms.

Bioavailability

We now need to consider Levine's work on *"bioavailability"*. Levine estimated that the bioavailability of vitamin C was complete with a dose of 200mg, but not at higher doses. But what is bioavailability and what does it imply? It sounds like an important biological parameter; the name suggests biological availability or usefulness within the body. Judging by the name, we might guess that the higher the bioavailability, the better for the organism.

In fact, bioavailability is nothing more than a measure of absorption from the gut. It is defined as the relative amount in the plasma obtained from an oral dose compared to an equal dose administered by intravenous injection (total oral dose plasma level / total injected dose plasma level x 100%). For example, suppose that by taking a one gram oral dose, you only get half the amount of vitamin C in your blood that you would get from a one gram intravenous dose: the bioavailability of the oral dose would be 50%. Essentially, only half the oral dose has reached the blood, the rest has been excreted. Despite the impressive name, bioavailability is at best a reflection of the absorption and excretion of oral doses of vitamin C.

Fred Hoyle pointed out that the establishment defends itself by complicating everything to the point of incomprehensibility. With Levine's evaluation of "bioavailability", the situation is worse than this, as besides being hard to understand, the measure is also biased. The bioavailability idea is taken from drug kinetics and is based on assumptions that do not apply in this context. Larger doses of vitamin C will give a lower "bioavailability", even when there is more vitamin C in the blood. Rather than being a measure of fundamental biological importance, it is misleading.

In Levine's study, higher oral doses of vitamin C resulted in greater levels in the blood. Furthermore, the plasma levels used to measure "bioavailability" were generally larger with increasing doses. However, compared to the equivalent injected dose, there was proportionately less. To simplify matters, with a low oral dose, all of it would be absorbed and available for use in the blood plasma; this is expressed as a bioavailability of 100%. If the oral dose were say, ten times larger, then maybe only half the administered dose (but still five times the low dose) would reach the tissues: this is expressed as a bioavailability of 50%. In this example, the absolute blood measure for the large dose is five times larger than that of the small dose, but the "bioavailability" of the large dose is only half that of the smaller dose.

Because bioavailability is a *relative* measure of the amount of vitamin C in plasma, the term is biased to give higher values at low doses. When Levine says bioavailability is complete at 200mg and incomplete at doses above this, it does not mean there is more vitamin C available with

200mg than with higher doses. Nor does it mean that an equivalent amount of vitamin C is available at the lower dose. Since the important question is the *absolute* availability of ascorbate in the tissues, the use of this relative measure is deceptive. The use of relative measures, such as percentage risk, was discussed earlier as a way of making small drug effects seem large. Levine's "bioavailability" compares each measured value with a different injected measurement. We can, however, easily remove the source of the bias and look at Levine's measured values. The results are:

Mean tissue availability of vitamin C			
Oral dose (mg)	200	500	1250
Actual tissue availability	24.72	34.05	54.58
"Bioavailability"	112	73	49

As we can see, the larger doses provide increased vitamin C to the tissues. (The tissue availability values here are mean areas under the time curves for Levine's measurements in plasma.) The important question is: do higher doses provide more vitamin C where it is needed? The answer is yes.

A picture might be useful to clarify the point. The next graph shows how, in a single individual, the plasma levels increase with time after a single oral dose. Notice how the base level is not zero but about 67 μM; below this value ascorbate is actively retained by the kidney. The chart demonstrates how blood levels increase substantially with the larger dose and how higher values are excreted more quickly. The result shows that a 1250mg dose raises blood levels more than a 200mg dose for the first six hours. Larger doses provide a bigger increase over this period. For the second six hour period, these and higher doses give similar blood levels.

Ascorbate plasma levels

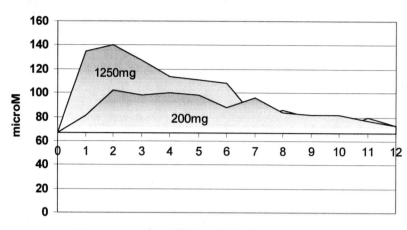

Hours since dose

Biological variation and dose

A feature of Levine's results is the amount of variation in the response of the subjects. This variation is consistent with Linus Pauling's claim that people vary widely in their requirements for vitamin C.

The Levine study did not use a high enough maximum dose of vitamin C to give a complete picture. It does show, however, that a substantial proportion of people will need doses of at least 2.5 grams for maximum effect. Levine has shown the amount of vitamin C in body fluids, or plasma concentration, rose as the dose increased, over the measured range from 30mg to 2500mg. In terms of blood levels, Levine's recommended 200mg dose (or even his 400mg "maximum") did not give the highest values in the study. In fact, based on the evidence presented in the paper, it is difficult to see how he reached his conclusions. In all cases, if the subject was given a higher dose, the plasma concentration increased. Only three male subjects were measured at both vitamin C doses of 400mg and 2,500mg and their plasma concentrations increased (from 77.3, 70.5 and 62 µM at 400mg to 84.7, 91.8 and 78.6 µM at 2.5 grams: increases of 9%, 30% and 37% respectively). Hence, even considering his own experimental results, Levine's recommendations for the RDA are on shaky foundations.

Levine has not refuted the suggestion that much higher values of vitamin C are needed in the diet. Furthermore, even within this relatively homogenous sample, he confirmed the variability in response to vitamin C. Given such variation in a small sample of selected young, healthy subjects, the recommended dose would need to be much larger for a large human population. Based on this data, the minimum recommendation that could be made to ensure everyone approached "plasma saturation" is 2,500mg - the largest dose actually studied. The requirement for many individuals would be considerably greater.

Raising blood levels

Despite our criticisms of the limitations of the data, Mark Levine should be congratulated for attempting to bring a more scientific approach to the recommended daily allowance. In discussion, Levine told us that he knows of no nutritional benefit from raising blood levels of vitamin C. Such evidence does however exist. For example, higher levels of vitamin C can be protective against damage to blood vessels,[83] and greatly reduce death rates in the elderly.[84] We can take from this that Levine is not convinced by the body of research linking higher intakes of vitamin C to good health. Such assumptions may have prevented Mark Levine from reaching arguably the most important conclusion from his data.

What Levine and others have shown is that vitamin C has a short half-life in the blood. Taking an oral dose will raise blood levels for only a few hours. Since the pharmacological properties of vitamin C are dose related, the benefit of a single dose is short lived. If high levels of vitamin C provide protection against the common cold, then a single oral multi-gram dose of vitamin C would have little more efficacy than a 500mg dose, as they raise blood levels in a similar fashion. The blood level would increase for a few hours and would return close to background levels for the greater part of the day. This interpretation of the biochemical results is clear and unequivocal.

At first sight, this seems to imply that Mark Levine has shown that Linus Pauling and others were wrong in suggesting megadoses of vitamin C to prevent disease. However, a single one-gram dose of vitamin C is not equivalent to the situation in animals which manufacture their own ascorbate internally, or who get the vitamin from food. Vitamin C could be manufactured continuously or be absorbed slowly from vegetable matter, being released gradually throughout the day as digestion proceeds. A vitamin C tablet would not raise blood levels in the same way. Indeed, five 100mg doses taken at intervals through the day would raise average blood levels more than a single one-gram dose.

The following graph shows how two doses of one gram each, six hours apart, raise levels in an individual's blood plasma. Notice how multiple doses will increase the mean blood levels and reduce the period of time when blood levels are near the minimum.

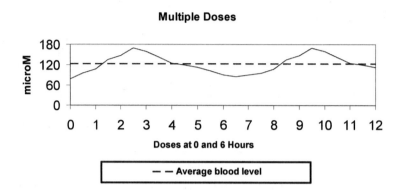

An individual who wanted protection from, say, the common cold by taking vitamin C, would raise their blood levels more effectively by taking divided doses or slow release formulations. This has immediate implications for studies that have tried to assess the effectiveness of ascorbate supplementation in the prevention of colds and other diseases. If a single dose of vitamin C raises blood levels for about six hours or one quarter of the day, the subject is unprotected for the other three quarters of the time. Studies of vitamin C supplements have often given conflicting results and much of the variation may be related to how the doses were given.

Levine's results indicate that Linus Pauling did not pay enough attention to how a supplement is given. They also provide a reason to suspect that the health benefits found with vitamin C supplements have been seriously underestimated. Some studies will have given negative results because of the timing of the supplementation.

Our reanalysis of Levine's data suggests that an optimal dose will be considerably greater than the RDA. The biochemical data supports Pauling's hypothesis that, for a large proportion of the population, the optimal daily dose of vitamin C is several grams per day. Levine's results draw attention to the importance of divided doses or slow release formulations of vitamin C. A single megadose tablet will only raise blood levels for a short period and is likely to be therapeutically ineffective. The aim is to raise plasma levels consistently and this requires either multiple tablets, taken at short intervals throughout the day, or the use of slow release formulations

Expensive urine

Levine's argument that excreted vitamin C serves no function is an oversimplification, a variation on the amusing but unscientific "expensive urine" jibe. A dose of vitamin C will increase average blood plasma levels before it is excreted, which could serve a multitude of purposes. With higher doses, some of it is retained in the gut and never enters the body. Vitamin C is expected to be a beneficial antioxidant in the gut and absorption may not be necessary for protection against stomach or colon cancer.[85,86] The ascorbate that enters the body passes through and is excreted in the urine. Importantly, animals that manufacture large amounts of ascorbate internally also excrete it in the urine.[87] Animals such as the rat excrete ascorbate, despite having used vital and often scarce energy to produce it. This implies either an impairment of evolutionary fitness or a substantial advantage to ascorbate excretion. Rats are vigorous mammals with a widespread population, which supports the second interpretation.

So how much do I need?

Despite all the hype, no one knows how much an individual human being requires for good health. Many have opinions and suggest that they have the answer, but the question remains open. We know for sure that people need a few milligrams each day to avoid scurvy. The idea that more is needed for good health is increasingly accepted. The actual minimum daily requirement for good health could vary over the range of, say, 100mg to over 20 grams, depending on the individual. Moreover, this is a difficult question to answer scientifically. We know of no simple experiment or practical study that could rigorously provide the answer.

The suggestions for the RDA are based on unscientific assumptions by establishment nutritionists. Despite Mark Levine's efforts to give them a solid basis in measurement, the values are not justifiable in terms of biology, statistics or basic science. It is simply not sensible to recommend a single value for a varied population. Scientifically, the arguments for the recommended daily allowance are barely justifiable as tentative suggestions. People who suggest that there is no compelling scientific evidence for megadose levels of vitamin C are invited to read the published justification for the RDA[63] by the Institute of Medicine critically. The establishment is in no position to cast aspersions.

Suggestions for megadose levels are also hypotheses; the evidence provided to support larger intakes is incomplete. Linus Pauling's orthomolecular approach suggests that people vary considerably and their required intake of vitamin C changes with their state of health. This is more biologically plausible than setting a fixed requirement for the entire population. The question boils down to how people deviate in their vitamin C requirements. If the variation is large, and the evidence suggests that it is, then Pauling is correct and a proportion of people need higher doses.

Is vitamin C safe?

"The tragedy of science is the slaying of a beautiful hypothesis by an ugly fact." T H Huxley

We now come to the tenuous evidence on the risk of toxicity. Low doses are clearly indispensable, since vitamin C is an essential part of the human diet. In the longer term, high doses may also be a requirement to avoid disease. The potential risk increases with higher doses but the level of vitamin C intake at which the benefits are outweighed by the hazard has yet to be determined.

Vitamin C is in the class of substances generally regarded as safe. No substance is completely safe, not even water; you can kill yourself by drinking too much water. Recently, it was reported that two parents in Springville, Utah, were accused of child abuse homicide for feeding their four-year-old daughter a large quantity of water.[88] Despite the widespread belief that water is safe, an excess intake can lower the concentration of sodium in the blood and cause the brain to swell. To put the safety issue into context, it would be easier to commit suicide by overdosing on pure water than by eating too much vitamin C.

A simple measure of the safety of a drug is the *therapeutic index*. This is calculated as the toxic dose divided by the therapeutic dose. The toxic dose is normally estimated to be the LD50, which is defined as the dose at which 50% of the subjects would die (Lethal Dose 50%). For example, a substance with a therapeutic dose of one gram per day, and an LD50 of two grams per day, would have a therapeutic index of two. Such a low therapeutic index indicates a dangerous drug, with little margin of safety. This kind of drug would be suitable for treatment in only the most severe life threatening conditions. One gram of vitamin C given to a 70 Kg man has a therapeutic index of at least 350, which is consistent with a substance generally regarded as safe.

A man trying to kill himself with vitamin C would need to take considerably more than 350 grams to have a reasonable chance of success. This large dose would need to be injected rapidly and intravenously, as a dose of this size taken orally would not be absorbed from the gut and would therefore be safe. Dr Robert Cathcart, one of the most experienced physicians in the field of high dose ascorbate therapy, states that the margin of safety for massive doses is much greater than that for aspirin, antihistamines, antibiotics, all pain medications, muscle relaxants, tranquillisers, sedatives and diuretics. In other words, vitamin C is far safer than all common drugs. By way of comparison, the painkiller paracetamol, also called acetaminophen, is widely available to the public and has a therapeutic index of about 25. Serious liver damage can occur with paracetamol doses as low as four grams, overdoses as low as seven grams are considered serious, and 15-gram overdoses are often fatal. (It is interesting that the damage is related to an oxidation process, which can be prevented using dietary antioxidants.[89,90,91]) Paracetamol is the most frequent cause of acute liver failure in the US.[92] This rather dangerous drug is widely available to the public and recommended for use in low-grade pain such as a mild headache.

We have not found a single report of a healthy person dying from a vitamin C overdose. By way of comparison, aspirin and the related non-steroidal anti-inflammatory drugs kill many more people every single year, than the number in the world who have ever died through taking vitamins. In the United States, at least 16,500 people die each year through taking anti-inflammatories and over 100,000 may be hospitalised by their side effects.[93,94] Again, in the United States, medical mistakes reportedly kill at least 100,000 people a year.[95] Twelve thousand cases of unnecessary surgery, 7000 errors in giving drugs, 80,000 hospital infections, 106,000 adverse drug reactions and 20,000 other errors, all lead to avoidable death.[96,97,98] Medicine may be the leading cause of death. By contrast, it is difficult to find *any* reliable reports of deaths from vitamins in the medical or scientific literature. Mark Levine states that "Harmful effects have been mistakenly attributed to vitamin C, including hypoglycaemia, rebound scurvy, infertility, mutagenesis, and destruction of vitamin B12. Health professionals should recognize that vitamin C does not produce these effects." [99] He was referring to the vitamin C scare stories that occur sporadically, producing medical myths that are accepted but untrue.[100]

Kidney stones?

The idea that vitamin C causes kidney stones is an old scare story that raised its head in the medical attack on Linus Pauling. It was a reasonable hypothesis, but unexpected kidney stones are not found in people taking large amounts of vitamin C. Indeed, vitamin C has even been proposed and used as a treatment for kidney stones.[101]

Mark Levine suggested that the possibility of kidney stone formation was a reason to limit the daily intake of vitamin C, although he appears to accept the absence of tangible data linking higher doses to stone formation. He argued that there was anecdotal evidence that oxalate stones might occur in response to higher doses, and added that no matter how small the risk, there would be some people in a sufficiently large population that would be affected. Despite the lack of evidence, Levine argued that oxalate and urate excretion might increase in relation to vitamin C dose. People with recurrent stone formation may have an unusual biochemistry, leading to an enlarged production of oxalate from vitamin C.[102] Oxalate and urate can accumulate in the formation of kidney stones. In practice, Levine found an increase in the excretion of both oxalate and urate with a dose of one gram of ascorbate, but earlier studies had produced variable results. He argued that if high doses were administered for longer, the increased oxalate excretion could continue and cause kidney stones. This theoretical side effect has not been observed.

Around three quarters of all kidney stones are composed of calcium oxalate; unlike some other stone types, these can form in acidic urine. Although vitamin C does increase the production of oxalate in the body, there is no evidence that it increases stone formation. It could even have the reverse effect, since vitamin C itself tends to bind calcium, decreasing its availability for formation of calcium oxalate. Vitamin C has a diuretic action: it increases urine flow providing an environment that is less suitable for formation of kidney stones. In addition, stone formation appears to occur around a nucleus of infection. High concentrations of vitamin C are bactericidal and might prevent stone formation by removing the bacteria around which stones form.

In a recent, large-scale, prospective study, 85,557 women were followed for 14 years; the study produced no evidence that vitamin C causes kidney stones.[103] There was no difference in the occurrence of stones between people taking less than 250 milligrams per day and those taking 1.5 grams or more. This study was a follow up of an earlier study on 45,251 men; this earlier study indicated that the doses of vitamin C above 1.5 grams reduced the risk of kidney stones.[104] The authors of these large studies stated that restriction of higher doses of vitamin C because of the possibility of kidney stones is unwarranted.

Large-scale studies have a role in searching for side effects of a treatment. If the incidence of side effects is low, then checking a large number of subjects is essential. However, like most large-scale studies, these kidney stone investigations were relatively unsophisticated in their implementation. It is difficult to give each subject more than a limited amount of time in investigation and data collection, when you have 50,000 or more to examine.

Vitamin C could prevent some types of kidney stones. Less common forms of kidney stone include uric acid stones (8%), that form in gout, and cystine stones (1%), which can occasionally be formed in children with a hereditary condition; these stones are not side effects of vitamin C. Other stones include those made from calcium phosphate (5%), which dissolve in a vitamin C solution. Acid urine produced by ascorbate will also dissolve struvite stones (magnesium ammonium phosphate) that often occur in infected urine. Whenever we read an uncritical warning of the risk of kidney stones in a document about vitamin C, we become suspicious that the author has not read the literature.

Side effect or useful property?

There is only one undisputed "side effect" for high doses of vitamin C, diarrhoea. The dose needed to produce this effect is variable. The level at which diarrhoea occurs is called the *bowel tolerance level*, and this can be used to indicate the amounts needed for supplementation. A single large dose of vitamin C can act as a natural laxative in the same way as a bowl of prunes, and provides an alternative to over-the-counter remedies for constipation. The medical establishment classes this action of vitamin C on the bowel as an adverse side effect, whereas prunes do not seem to have a recommended upper limit. Presumably, people are expected to use their common sense when they eat prunes, but not when supplementing with vitamin C.

Unbelievably, together with the RDA, a tolerable upper intake level has been set, based on loose stools.[64] The US Institute of Medicine could find no other side effects of large doses of vitamin C in healthy people. They set the maximum intake at two grams of vitamin C per day, as at this level, almost no one in the population will suffer from diarrhoea. We tentatively suggest that if a dose causes diarrhoea, the subject might notice and reduce their intake.

The danger of too small a dose

The RDA committee do not know the extent of human variation in vitamin C requirements, and have the unrealistic task of setting a single figure for an upper limit that will apply to all. Basic numeracy suggests that this "recommended maximum level" is not optimal for the individual. They assume that all people, regardless of age or state of health, will be receiving adequate vitamin C at this dose. We know of no solid evidence to support the assumption that two grams per day will cover all human variation in the requirements for vitamin C.

The available evidence suggests that large doses of vitamin C are safe.[105] There are no double-blind clinical trials showing adverse effects from large amounts of vitamin C. Substantial experience suggests a high degree of safety.[106,107] In some medical conditions, such as kidney disease, haemochromatosis (a hereditary condition in which the body stores excess iron), and deficiency of the enzyme glucose-6-phosphatase, there are theoretical side effects of extremely large doses of ascorbate. People with these conditions would be wise to seek medical advice before considering high dose supplementation.[63] Such considerations do not apply to normal members of the population. Robert Cathcart questions the universal validity of these contraindications. He reports that he has tried but has been unable to trace the author of the paper on glucose-6-phosphatase deficiency. He suggests that intravenous ascorbic acid may have been used inappropriately instead of sodium ascorbate. He also reports experience of two patients with haemochromatosis that he treated with massive doses of ascorbate, without problem. In the thousands of patients he has treated, he has never seen any evidence of a damaging iron related reaction.

Those making official nutritional recommendations have concentrated on the dangers of overdosing with vitamin C. It is theoretically possible that one gram or more per day might cause harmful effects, but this has yet to be shown. However, the alternative hypothesis, that greater dangers could result from too low a recommended dose, is ignored.

Paradoxically, attempts to minimise the recommended dose based on hypothetical harmful effects may lead to increased disease. The evidence for the beneficial effects of larger doses, such as 500mg and above, is much stronger than the evidence for side effects. For a normal healthy individual, the risk of a 10-gram per day dose may be so small as to be inconsequential. Conversely, if the claimed benefits of large nutritional doses are even partly true, many millions of people could live longer, healthier lives and avoid serious diseases by supplementing their diet. According to this analysis, the risk of consuming high doses of vitamin C is far lower than that of not consuming enough.

Biased experiments

**"Truth is ever to be found in the simplicity, and not in the multiplicity and confusion of things."
Isaac Newton**

The therapeutic effects of vitamin C are dose dependent and, unlike many drugs that become toxic at higher doses, ascorbate remains safe. The majority of dietary antioxidant supplements are also harmless. However, because of the natural conservative bias in society, the media and medicine, it is easy to grab the headlines with a sensational vitamin C scare story. Such stories are almost guaranteed media attention far beyond their scientific merit or relevance. Here we describe some of the more recent and controversial reports.

Conference report: vitamin C clogs arteries

A report presented in March 2000, at the 40th Annual Conference of The American Heart Association, suggested that vitamin C in high doses clogs the arteries.[108] The authors, Dr Dwyer and colleagues, concluded, "Regular use of vitamin C supplements may promote early atherosclerosis". Despite this report originating from an oral presentation at a conference, as opposed to a well-regarded, refereed journal, Dwyer's alarming conclusion was given a high profile in the international media. According to the study, people who had taken supplements of vitamin C for 18 months had thicker carotid arteries: the measured increase in arterial wall thickness was reported to be 2.5 times that in the control subjects.

The measure 'increase in arterial thickness' is a relative value and, as we have seen, such measures are often used to magnify research results and make them seem more important. In this case, the value relates to the difference in wall thickness over the period. What Dwyer actually reported was a difference in the rate of increase in wall thickness. The actual change could be so small as to be irrelevant. If, say, the increase in thickness were 1% for the control group, then it would be only 2.5% for the vitamin C group. Since the absolute increase in the thickness of the wall was not reported, the claimed "2.5 fold increase" was meaningless.

Conflicting studies

A well-designed study, published by Stephen Kritchevsky in 1995, contradicts the results of Dwyer's misleading report.[109] Kritchevsky measured the thickness of the carotid artery wall in 6,318 female and 4,989 male subjects, aged from 45 to 64 years, and found a significant *reduction* in thickness in people over 55 who had consumed one gram or more of vitamin C each day. Considering other factors, higher intakes of vitamins C and E were found to be related to reduced artery wall thickening. Other clinical studies have found vitamin C to have beneficial effects on the relaxation and widening of arteries. For example, a recent study suggests that the vitamin is superior to diltiazem, a drug used for relaxation of blood vessels and treatment of angina.[110]

In response to Dwyer's presentation, the Life Extension Foundation asked Paul Wand, a neurologist, to investigate. Wand found reports suggesting that vitamin C was protective against atherosclerosis. Unfortunately, the low doses usually investigated did not provide information that could be extended to high dose users. He could find almost no information concerning doses greater than two grams of vitamin C per day, and people who take these high doses often supplement with other antioxidants, such as coenzyme Q10 and selenium. Wand carried out an extensive literature search on the effects of vitamin C on cardiovascular disease risk, using Medline. The search was from January 1, 1990 to April 25, 2000.

Wand's review showed that subjects taking more than 500mg of vitamin C per day were more likely to show a beneficial response. Doses below 500mg were not effective against cardiovascular disease.

Effects of vitamin C on cardiovascular disease (1990–2000)		
	Dose < 500mg	Dose > 500mg
Beneficial response	1 study	30 studies
No response	3 studies	4 studies

Dr Wand followed up his literature search by conducting a pilot study on 30 subjects, who had taken a minimum of two grams of vitamin C daily for at least four years. The people in this study also consumed other supplements. The age of the subjects varied from 45 to 81 years, with an average of 61 years. These subjects were therefore older than Dwyer's. Wand measured the carotid artery thickness of his subjects using high-resolution ultrasound and Doppler evaluation, which estimates the blood flow through the vessel. Multiple scans through the right and left carotid systems were used to see if plaque was present. This procedure allowed estimation of wall thickening, measurement of blood flow and of the degree of stenosis, or blockage. Dr Wand uses this technique as part of his normal clinical routine and in everyday practice. He regularly finds 60% to 90% blockage, together with severe wall thickening in the carotid arteries. As the artery becomes blocked, the speed of blood flow is normally faster; this increase in speed is expected from the physics of fluids.[111]

Dr Wand's pilot study showed that 23 out of 30 of these high vitamin C subjects displayed no evidence of plaque formation, blockage or wall thickening, and had normal blood flow. There was some pathology in the remaining seven subjects. In five, the disease was slight and considered insignificant, while the remaining two subjects had a degree of carotid blockage estimated at 30% and 40%. As these two subjects were of advanced age, this degree of blockage was considered normal. The seven cases with indications of carotid pathology had high levels of homocysteine, LDL cholesterol or glucose, which may have contributed to the findings. Wand reported that problems with these seven subjects were less than expected, given these other contributory factors. Overall, Wand's group of vitamin C supplement takers had remarkably healthy arteries. The results of Wand's study are consistent with the hypothesis that vitamin C can prevent or even reverse atherosclerosis.

Even if we take the Dwyer study at face value, it provides no evidence that Vitamin C blocks carotid arteries. The changes reported by Dwyer could have been a reversal of the normal, age-related thinning of artery walls, which would not imply blockage but a strengthening of the vessel. Since blood flow was not measured, we do not have sufficient information from Dwyer to determine whether the change was detrimental or beneficial. Over two years after the conference presentation, the Dwyer study had still not been published and was described as "under review".[112] The study by Dwyer was eventually published and claimed some antioxidants were helpful in preventing atherosclerosis but did not provide data to support the negative claims for vitamin C.[113] Dr Dwyer declined our request for additional evidence or explanation.[112]

More recently, a well-funded multidisciplinary study found no significant therapeutic effect of vitamin C (one gram) and vitamin E (800 IU) supplements in postmenopausal women with a pre-existing degree of blockage of the coronary artery,[114] over a period of 2.8 years. In this randomised, controlled trial, researchers measured the arterial wall thickness using angiograms, a type of x-ray allowing visualisation of the flow of blood through arteries in the heart. This is an additional non-response study to include with those found by Dr Wand in his literature review. This makes five studies that showed no benefits of vitamin C (over 500mg), while 31 (including Wand's own) gave positive results.

Heart Protection Study: vitamins a waste of money?

A recent, large-scale study claimed that antioxidant vitamins are not beneficial for heart disease.[115] The research was carried out by the prestigious UK Heart Protection Study Collaborative Group of researchers and hospitals. This study was reported on television news and on the front page of national newspapers. It was suggested that antioxidant vitamins are a waste of money.

The Heart Protection Group study cost well over 20 million dollars and involved 20,536 adults, with ages from 40 to 80 years. These subjects had coronary disease, other occlusive arterial disease, or diabetes. The patients were randomly allocated to groups receiving either a combination of 600mg vitamin E, 250mg vitamin C and 20mg of beta-carotene daily (treatment group) or a matching placebo (control group). The study lasted five years and most participants remained in the study over this period. In the treatment group, blood levels of vitamin E doubled, levels of vitamin C increased by one-third, and levels of beta-carotene quadrupled. The vitamin supplements did not alter the number of deaths, heart attacks, or strokes. The conclusion was that while the supplements appeared to do no harm, they did no good either.

Statins versus antioxidants

Drug companies do not waste millions of dollars of their shareholders' money funding large-scale studies on vitamins that offer no potential for profit. In the case of the Heart Protection Group study, additional results were announced concerning simvastatin, which is a cholesterol-lowering drug or *statin*. The report claimed that adding simvastatin to existing treatments conferred substantial benefits on high-risk patients, regardless of their initial blood cholesterol levels. The doses of statin were reported to be safe; however, significant side effects have been reported with these drugs, such as a potential cancer risk.[116] Statins are drugs that inhibit the synthesis of cholesterol, which, contrary to popular belief, is an essential part of human biochemistry. Statins also lower the levels of an essential antioxidant, coenzyme Q10, and may have long term side effects such as heart failure, cancer, Parkinson's disease and cataracts.[117,118]

Heart failure could increase with exposure to statins. Pharmaceutical companies are aware of these risks and hold patents on formulations of statins combined with nutritional supplements, such as coenzyme Q10[119,120] or carnitine,[121] to prevent heart failure and other severe side effects. Being aware of the potential for statins to cause heart failure and muscle problems, presumably the companies will add coenzyme Q10 to their products when the side effects are "proven". Dr Julian Whitaker has petitioned the FDA that statin drugs should carry a label recommending the use of coenzyme Q10 with these drugs, to avoid damage to heart muscle in the longer term.[122]

The Heart Protection Group reported that a daily simvastatin dose of 40mg reduced heart attacks and strokes by about a quarter. They concluded that, in high-risk individuals over five years, simvastatin would prevent 70 to 100 people in every thousand suffering one of these major life-threatening events. People with low blood cholesterol would also achieve these benefits. It was additionally reported that if the findings of these studies were incorporated into clinical practice, the drug companies hoped to benefit by billions of dollars worth of sales of statin drugs. The implications of the Heart Protection Group's reports are clear: that everyone should be taking one of these expensive statin drugs, that over many years the drug companies will benefit, and that vitamin pills are a waste of money and should be thrown away.

How to design a bad experiment

We believe that the Heart Protection Group study described earlier is biased and its conclusion, that supplements are not beneficial to heart disease, is misleading. This study included a large number of patients, which makes it seem impressive but restricts replication. The study extends over a number of years and while this also gives the appearance of scientific rigour, it increases the cost. Few doctors have 20 million dollars available to repeat this work if they believe it to be biased.

Using the wrong kind of vitamin E

The Heart Protection Group results on vitamin E are biased for a number of reasons. To begin with, the study used synthetic vitamin E (dl-alpha-tocopherol), which is less biologically active than the natural form. The name vitamin E does not refer to a single chemical, but is a name for a mixture of chemicals. The most common of these are the tocopherols, but the related tocotrienols also have vitamin E activity in the body. The tocopherols exist in a number of forms called alpha-tocopherol, beta-tocopherol, gamma-tocopherol and delta-tocopherol. Each variety of tocopherol has its own biological properties. Alpha-tocopherol is generally considered a more powerful antioxidant than the other types and, in the past, was considered the preferred form.

Research suggests that while alpha-tocopherol is a more potent antioxidant, gamma-tocopherol may be linked more closely to heart disease.[123,124,125,126] Gamma-tocopherol can exert a more powerful inhibitory effect on inflammation,[127] which is involved in arterial plaque formation. A Swedish study found blood levels of gamma-tocopherol, but not alpha-tocopherol, were reduced in patients with heart disease,[128] while the alpha form is preferentially depleted in smokers.[129,130] Furthermore, supplementing rats with alpha-tocopherol alone can decrease blood levels of gamma-tocopherol.[131] The gamma form is likely to be an important part of many metabolic processes in the body but it has been harder to measure than alpha-tocopherol. Many health food stores recommend vitamin E tablets that contain mixed combinations of the tocopherols.

The use of synthetic vitamin E in the Heart protection Group Study would reduce the effectiveness of the antioxidants. In the early 1960's, it was noted that negative studies on vitamin E had generally used the synthetic form. There is essentially no difference between natural and synthetic vitamin C, but natural and synthetic vitamin E are very different. Organic molecules like alpha-tocopherol can exist in two forms. These types are called optical isomers, the d-form and the l-form, and they are chemically very similar. These different forms have the same chemical formula, contain the same number of atoms, and are connected together in the same sequence. The difference in their structure is like that of our left and right hands: identical but mirror images of each other. They are so similar that the way to tell them apart is by checking the direction in which they rotate a beam of polarised light.

For the vast majority of chemical reactions, the two forms are identical. The main exception to this rule is in biology. One of the characteristics of biological chemistry is that sometimes only one form of optical isomer can be used for a particular purpose. This is because the enzymes are very specific and generally react with only one form of a molecule. Enzymes react with molecules in a way that is described as a lock-and-key fit. The enzyme has an active site that the molecule fits into in order to react. Right hands will not fit neatly into left-handed gloves and, similarly, you often need the correct optical isomer to fit the enzyme. In this way, the body uses only one of the d- or the l- form of such component chemicals, but not both.

The synthetic vitamin E used in the statin study is expected to be biologically less effective. Synthetic vitamin E is made using standard chemistry and contains both the d- and the l- forms of alpha-tocopherol, whereas normally, only d-alpha-tocopherol is found in the body. Both d- and l- forms of alpha-tocopherol have the same antioxidant potential but do not have the same biological activity.[132,133,134] Theoretically, l-alpha-tocopherol is so similar to the d- form that it might interfere with some biological reactions. Supplementing pigs with natural vitamin E produces higher blood levels than the synthetic form.[135]

Synthetic vitamin E also contains seven unnatural forms that have a tail in an "S" configuration, with pronounced "kinks" that natural forms do not have. While the synthetic vitamin E molecules with "S" shaped tails do enter membranes, they do not stay there. The kinked tails twist out of the plane of the membrane and prevent the molecules stacking close together. Natural vitamin E molecules can pile together, rather like a set of spoons, and are more compatible with membrane lipids. Synthetic vitamin E is also poorly absorbed unless taken with a substantial amount of fat or oils. Cold-water dispersible forms of natural vitamin E are more easily absorbed from the diet and are recommended by health food suppliers.

Use of a nutritional dose

A second cause for concern is that the Heart Protection Group used too small a dose of vitamin E. As a nutritional dose of synthetic vitamin E, 600IU daily is reasonably large. Since this was a study of subjects with disease, a pharmacological dose of at least 1600IU of natural vitamin E would have been more appropriate.

Nutrition confused with pharmacology

The Heart Protection Group study confuses normal nutrition of the healthy with pharmacology for treating the sick. The vitamin doses they used would have little or no therapeutic effect. From the review by Dr Wand, we have seen that doses of vitamin C lower than 500mg are unlikely to be effective in preventing heart disease. By using 250mg, the study has stayed well below the doses that might produce a positive result. Furthermore, if vitamin C is in short supply, vitamin E may act as a pro-oxidant instead of an antioxidant. The Heart Protection Group knew that vitamin C is generally considered safe and might easily have used a pharmacological dose of, say, three grams per day in divided doses or sustained release formulations. However, they chose not to do this.

The difference between nutritional and pharmacological doses is a relatively sophisticated distinction, and few in the media or medical circles noted the error. Linus Pauling's therapy for prevention and treatment of heart disease involves six grams of vitamin C per day. Dr Cathcart would argue that a pharmacological dose of vitamin C is the gut tolerance level, which may be considerably greater than this, let us say as much as 25 grams daily in many patients. The Heart Protection Group's nutritional dose was less than 5% of the dose Pauling suggests, and 1% of the dose Cathcart might recommend.

Biased heart study

Following careful examination, the Heart Protection Group study no longer appears reliable and we conclude that the results of this study are inapplicable to high dose supplementation.

Despite initial appearances, it would have been hard to design a more misleading experiment than the one conducted by the Heart Protection Study Group. We accept that the doctors involved in the Heart Protection Study Group study were ethical and operated what they thought was a good, solid, scientific investigation, using their extensive knowledge of medicine in general and cardiovascular disease in particular. However, they missed the point entirely and spent a fortune producing a nonsensical and biased study. It is worth remembering that they made a strong public claim based on insufficient evidence. The researchers claimed their results showed antioxidant supplementation to be of no value; but scientifically, their study was useless in terms of determining the importance of high dose supplementation in heart disease.

Avoiding the issue?

Many researchers into vitamin C seem to avoid the issue of benefit deliberately. A recent study of vitamin supplements, by Dr Michael Gaziano and colleagues, concluded they offered no benefit for people with coronary heart disease. This prospective study was based on 83,639 subjects in the United States, who had no history of cardiovascular disease or cancer. There were 1037 deaths from cardiovascular disease, which included 608 deaths from coronary heart disease, over a period of five and a half years. The authors concluded, "In this large cohort of apparently healthy US male physicians, self-selected supplementation with vitamin E, vitamin C, or multivitamins was not associated with a significant decrease in total cardiovascular disease or coronary heart disease mortality".[136] This large study over an extended period was also misleading.

Dr Joel Simon from California read this paper and noticed that the conclusion was untrue – the results from one group of subjects suggested a large, significant benefit from vitamin C and E.[137] Low-risk subjects, taking both vitamin C and vitamin E supplements, showed a 41% reduction in deaths from coronary heart disease and 34% from cardiovascular disease. If these figures were correct, then 41 out of every 100 people in this group who died of a heart attack might have lived if they had taken supplements. Furthermore, 34 out of every 100 who died from cardiovascular disease might not have died. These are large percentages, indicating that supplementation could have a major benefit; however, the researchers ignored these results.

Dr Gaziano subsequently agreed with Dr Simon that his Physicians' Health Study results actually indicated a 28% to 41% reduction in mortality, with the greatest effect being in the subgroup taking both vitamins.[138] The results in question related to a group with low-risk factors. Gaziano claimed to have analysed this subset of subjects because previous observational studies had shown benefit in low-risk individuals, in contrast to results on high-risk groups. His results actually confirmed the prior studies, but the replication of these findings was ignored because they could have arisen by chance. This explanation is a little bizarre, as all statistical results are based on probability and their result was statistically significant (which means unlikely to occur by chance alone).

In his reply to Simon, Gaziano pointed out that the results indicate the possibility that vitamins C and E are effective in the early stages of atherosclerosis. In his own words, "…many observational studies among those at usual or low-risk of CVD suggest a benefit of these vitamins. These findings from our study as well as the conflicting data from trials and observational studies raise the possibility that vitamins E and C are most effective in the earliest stages of atherosclerosis."

This statement, that vitamins C and E might be effective in preventing the development of atherosclerosis, is the opposite of their original claim. The confusion arose, yet again, from the erroneous idea that experiments based on large numbers of subjects are more valid.

Science works so effectively because it is based on experiment and replication. In this case, we have a clear example of a single large experiment being misunderstood by the very people who conducted it. Fortunately, the error was pointed out and the researchers were then able to correct their bias and explain the position more accurately.

Oxidation and illness

"There are more than ten thousand published scientific papers that make it quite clear that there is not one body process (such as what goes on inside cells or tissues) and not one disease or syndrome (from the common cold to leprosy) that is not influenced - directly or indirectly - by vitamin C." Drs Emanuel Cheraskin, Marshall Ringsdorf and Emily Sisley

Biological mechanisms for controlling illness and fighting infection have been honed by millions of years of evolution. The role of vitamin C as an antioxidant is fundamental to these mechanisms. Ascorbate has a primary biological role in the resistance of both plants and animals to the harmful effects of free radicals, in many forms of stress and disease.

Oxidation, stress and disease

Oxidation causes damage in many forms of illness; infection, inflammation, shock and cell death all involve free radicals. For this reason, antioxidants have been proposed as a new approach to fighting disease.[139] Free radicals are also involved in the control of the body's reaction to insult. Throughout the plant and animal kingdoms, response to the stress of infection, damage or cancer depends on antioxidants and redox signalling. Redox signalling is when cells communicate using oxidants and antioxidants.

Oxidative stress plays a part in many diseases, although its relative importance as a disease mechanism is not established.[140] In many diseases, it may be the primary mechanism. In others, such as the action of slow viruses, the involvement of free radicals in the disease process may be minimal. The suggestion that free radical damage plays a substantial part in many infections is not controversial. Even diseases that have a specific and unrelated cause, such as diabetes, involve free radical damage.

Oxidation in diabetes

Diabetes mellitus is a common illness, resulting from a lack of the hormone insulin, which controls the distribution of glucose in the body.

The relationship between ascorbate and glucose is less well known. Vitamin C and glucose have similar molecular structures. They compete for the biochemical pumps that transport them into cells, so high levels of blood sugar restrict the amount of vitamin C entering the cells.

Since the work of Banting and Best in 1922, we have been able to control diabetes by injecting insulin.[141] The introduction of insulin as a treatment for diabetes has saved many lives in the short-term, but diabetics still suffer a number of long-term problems. Insulin therapy does not cure diabetes: treated diabetics survive but are prone to atherosclerosis, heart attack, stroke, diabetic retinopathy (vascular disease of the eye leading to blindness), cataracts, kidney failure and damage to the peripheral nerves.[143] These chronic symptoms are thought to result from the inability to control blood glucose levels properly with supplemented insulin.

The high levels of glucose found in diabetes competitively inhibit the uptake of vitamin C into cells, so symptoms such as diabetic atherosclerosis are associated with ascorbate deficiency.[142] Diabetics are under substantial free radical stress. They have raised levels of oxidised lipid in their blood and lower plasma levels of vitamin C.[143] In particular, glucose can damage proteins by forming molecular cross-links, in a process called glycation. The cross-linked proteins become additional sources of oxidation damage. Thus, we can see that even in a disease like diabetes, in which the primary cause is known and treated, free radical damage plays a substantial role.

Cell death

Oxidation plays a significant part in the death of cells. Illness and injury to the body causes cells to die by two mechanisms, called apoptosis and necrosis. In apoptosis, a cell activates its own "suicide pill" and destroys itself. Apoptosis, the process of programmed cell death, involves free radicals.[144] Cells dying in this way release large amounts of oxidants into their surroundings.[145] Oxidants act as signals to initiate and control programmed cell death, which may be slowed or even prevented by the addition of antioxidants.[36,77,78] In necrotic cell death, cells swell up and rupture. The content of the cell is released into its surroundings generating additional free radicals. Classification of cell death into apoptosis and necrosis is not absolute and, in many cases, the death of cells includes elements of both.

Inflammation

For a long time, inflammation was a side issue in medicine and was viewed primarily as part of the body's defence against infection. Its role in the development and progression of disease was underestimated. Inflammation occurs when white blood cells, which provide the first line of defence against infection, invade a tissue and become active. The widespread nature of inflammation means that we all have some personal understanding of the process. Nearly everyone has experienced the irritation of an insect bite or a pimple on the skin. By understanding inflammation, we begin to appreciate how large doses of vitamin C can influence the progression of disease. Once again, free radicals play a fundamental role.

Vitamin C and inflammation

Diagram: Infection, Heat shock, Allergy, Local damage, and Trauma all lead to Oxidation, which causes Inflammation. A cycle between Dehydroascorbate and Ascorbate provides Redox control over inflammation.

Inflammation is thought to cause of about 30% of cancers and is associated with free radical damage.[146,147,148] A classic example is asbestosis, which starts when microscopic particles of asbestos are breathed into the lungs. The jagged particles irritate the surrounding tissue, causing white blood cells to converge. The tissue becomes inflamed as the white blood cells secrete free radicals, in an attempt to destroy the foreign invader. The tough, inorganic asbestos particles are undamaged by the onslaught, but the free radicals damage the surrounding cells, leading ultimately to cancer. The cancer is produced by the body's response to the asbestos particle, namely inflammation and free radicals, rather than by the particle itself.

At the start of an inflammatory process, small blood vessels near the area of damage expand and increase blood flow. Later, the flow of blood slows. Endothelial cells that line the blood vessels swell and no longer form an intact internal lining. As the lining of the blood vessels becomes incomplete, they become leaky and allow water, salts, and some small proteins to pass from the blood into the damaged area. This fluid seeping from the blood vessels causes local swelling. White blood cells, especially neutrophils, stick to the swollen lining of the blood vessels and then actively migrate through the walls into the area of tissue damage. Small white blood cells, macrophages and lymphocytes, follow the neutrophils. Macrophages ingest bacteria and other foreign matter, while lymphocytes secrete local hormones that stimulate inflammation.

The activation of white blood cells is controlled by oxidants and depends on redox signalling.[149,150] White blood cell activation releases free radicals into the tissue.[36,151,154] Activated neutrophils and macrophages take up oxygen and consume glucose. Indeed, the amount of additional oxygen used can be up to 20 times higher than normal.[36] This respiratory burst is used for the production of oxidants. White blood cells maintain high intracellular vitamin C levels[152,153] to protect the cell body from the oxidative side effects of activation while supporting the core defensive process.[156,154] As activated white blood cells boost their metabolism, antioxidant recycling of vitamin C increases to as much as 30 times normal levels.[155]

Phagocytes

Phagocytes are cells that eat bacteria and other unwanted particles. Amoeba-like macrophages and neutrophil white blood cells can engulf solid particles, such as bacteria, and destroy them with oxidants. These cells are the primary method used by the body to remove foreign microorganisms.

As inflammation becomes established, monocyte white blood cells are attracted into the damaged area. These white blood cells change into actively phagocytic macrophage cells when they enter inflamed tissue. Some macrophages are found in normal, healthy tissue, but they concentrate at sites of inflammation and damage. Macrophages engulf solid particles and dead cells, as well as bacteria and viruses, and can remove unwanted or necrotic material. Vitamin C protects these white blood cells from oxidation damage, without interfering with their ability to engulf foreign particles.[156] Ascorbate is therefore protective in conditions involving inflammation and white blood cell activity.[36,157,158,159,160,161,162,163,164]

Killing bacteria by oxidation

Phagocytes and other white blood cells destroy bacteria using free radicals and oxidants.[165] Their respiratory burst produces free radicals[166] in a complicated process.[167] Neutrophil white blood cells can still destroy some bacteria if the respiratory burst is prevented, but the normal killing process for phagocytes is dominated by the production and use of free radicals.

External release of free radicals allows phagocytic cells to attack invaders that are too large to engulf, such as parasitic worms. Many other non-phagocytic cells, such as glial brain cells and osteoclasts in bone, can also generate the release of superoxide in a respiratory burst. However, releasing oxidants into the surrounding fluid may cause damage to nearby tissues.[5,169] In acute, local inflammation, this damage may be small. In chronic or widespread inflammation, however, the injury may be significant.

The biochemistry of superoxide production is largely understood.[36,165,166] Inside the phagocyte, bacteria wrapped in a bubble of cell membrane are subjected to a high concentration of superoxide. The engulfed particle, together with the membrane that surrounds it, is called a vacuole. It appears that the inside of the vacuole is made acidic, whereupon the superoxide combines with protons to form hydrogen peroxide. Hydrogen peroxide, H_2O_2, is capable of killing many forms of bacteria, as is well known from its use as an antiseptic.

The green colour of pus, common in infected wounds, comes from the presence of another oxidant, myeloperoxidase. Myeloperoxidase is a non-specific oxidant, which can kill bacteria and fungi in the presence of hydrogen peroxide and chloride ions. This is an enzyme present in neutrophils but not macrophages, and was first isolated from large volumes of human pus.[168] The enzyme is found in granules within the neutrophils that are fused with the engulfed vacuole. This enzyme may use chloride ions to form the highly reactive hypochlorous acid, HOCl, which can oxidise many biological molecules and will decompose to poisonous chlorine gas.[169]

Cell damage and oxidation

The energy producing mitochondria in healthy cells release low levels of superoxide but those in damaged cells produce much more. In slices of rat lung exposed to air, production of superoxide increases from the normal level of about two percent, to nine percent.[43] In higher concentrations of oxygen, as much as 18% of total oxygen uptake is converted to superoxide. Mitochondria in ischemic tissues, deprived of oxygen, also produce more free radicals than those in normal cells.[170] Injury puts damaged cells under oxidative stress.

Anti-inflammatory substances

Many antioxidants prevent inflammation. Inflammation involves a number of different cell types, chemical mediators and local hormones. Perhaps the most well known inflammatory chemical is histamine; antihistamine creams are widely used to minimise irritation. More interesting is the involvement of other local hormones, the best known of which are the prostaglandins. Prostaglandins act locally and come in different forms. Some prostaglandins promote inflammation while others inhibit it. Vitamin C modifies prostaglandin synthesis and this explains some of its effects in inflammation, expanding blood vessels and reducing blood clotting.[171]

Many people deal with inflammation by taking analgesic painkillers. Aspirin and paracetamol act by inhibiting prostaglandins. In this case, the drugs block the enzyme cyclooxygenase, or COX, that is involved in the synthesis of prostaglandins. Unfortunately, aspirin blocks the formation of two types of prostaglandin, those that prevent inflammation and those that enhance it. The COX enzyme exists in two forms, called COX_1 and COX_2. The COX_1 enzyme produces the prostaglandins that prevent inflammation and the COX_2 enzymes those that enhance it. (Paracetamol may act on a third form, called COX_3[172] but this awaits verification.)

Some drugs are more specific than aspirin and affect only COX_2, reducing the inflammatory prostaglandins without affecting the others. These more specific drugs are generally expensive. However, curcumin, a derivative of the herb turmeric that gives curry its yellow colour, is a potent inhibitor of COX_2 but does not affect COX_1. It has the benefits of being both an antioxidant and a potent anti-inflammatory; it is relatively cheap and tastes good. So why do we not hear about its properties more often?

In 1995, two scientists from the University of Mississippi obtained a US patent on the use of turmeric for healing wounds. To obtain a patent, you need to demonstrate that an idea is new. The application accepted that turmeric had long been used in India as a traditional medicine for treatment of various sprains and inflammations. They claimed, however, that there was no research on the use of this powerful antioxidant as a treatment for external wounds. The Indian government challenged the patent. India provided evidence that turmeric had been used locally to treat wounds for many years. Ultimately, the patent application was rejected. Ironically, the failure to obtain a patent, and hence protect potential profits, may be the reason that few people in the West know the healing power of turmeric.

Antioxidants and inflammation

The links between inflammation and oxidation are clear: free radicals activate the inflammatory response and an inflammatory response produces oxidising free radicals. Infection by bacteria or viruses results in consistent free radical damage.

The presence of free radicals activates white blood cells. Neutrophils activate as needed, since they recognise dehydroascorbate and other oxidants, which accumulate with injury. These white blood cells take up the dehydroascorbate and reduce it back to ascorbate. By taking up the oxidised dehydroascorbate, rather than the reduced ascorbate, neutrophils conserve the local availability of vitamin C. The main danger for the activated neutrophil appears to be free radical damage to lipid cell membranes. The cell membranes are protected by vitamin E but the small amount in the cell walls can soon be oxidised, if not regenerated by ascorbate. Vitamin C reduces oxidised vitamin E, allowing it to continue to provide protection.[173] This reduction of vitamin E is a further reason why white blood cells accumulate so much ascorbate inside their cell bodies.

The levels of ascorbate required in disease may have been grossly underestimated. By controlling oxidation, vitamin C changes the body's response to disease. In the short term, it may prevent an acute overreaction, leading potentially to shock and death. In the longer term, it prevents chronic inflammation.

The ultimate antioxidant

"It is my hypothesis that what makes ascorbate truly unique is that very large amounts can act as a non rate-limited antioxidant free radical scavenger." Robert F. Cathcart

Vitamin C alters the body's response to disease. In 1985, Dr Robert Cathcart developed a theory of ascorbate as the ultimate biological antioxidant.[14] Although other dietary supplements, such as grape seed extract or lipoic acid, are stronger antioxidants, their maximum concentration in the body is more limited. Vitamin C is water-soluble and can be given in very high doses, resulting in a flow of molecules through the tissues. As the ascorbate in a tissue is used up, it can be replenished from the blood stream. This continuous flow supplies the electrons needed to prevent cell damage.

"Megadoses" are small

We start from the premise that many of the ideas about what are conventionally termed megadoses of vitamin C are wrong. Such doses, in the region of one to several grams, are simply not large enough! These megadoses may be important in maintaining health and preventing disease but are inadequate for treatment of acute illness. An effective therapeutic dose may be 1000 times that suggested for nutrition, a difference of three orders of magnitude.

In science, differences of this size often imply new properties.[174] For example, a person asked to walk very slowly might progress at one mile per hour. However, if we speed the person up by a factor of 1000, the accelerated walker would be beyond the sound barrier. The emergent properties are quite different. If we introduce a brick wall as an obstruction, the slow person would simply come to a halt, gently pushing up against the brickwork. The accelerated person, in contrast, would hit the wall in an explosion that would destroy both. The difference between Cathcart's suggested therapeutic doses and the RDA are greater than the difference in speed between walking and supersonic jets. The magnitude of this difference is why most studies on vitamin C have little relevance to the use of massive doses for the treatment of disease.

Cathcart's theory is an extension of an idea that originated with Dr Fred Klenner. In 1971, Klenner reported the use of large doses of vitamin C as a drug.[175] His report to the Journal of Applied Nutrition has an editor's note, pointing out that he really did mean doses of 150 grams per day, given intravenously. This large dose was so unusual that the editor of the journal had to indicate that it was not a misprint. Klenner observed that the use of ascorbic acid in human pathology follows the *law of mass action*, which states that, in reversible reactions, the extent of chemical change is proportional to the active masses of the interacting substances. In other words, the more vitamin C, the greater is the effect.

A limited supply of antioxidants

We might wonder why only vitamin C is able to quench large numbers of free radicals in this way. Normal antioxidants, such as vitamin E, are limited in their ability to reduce free radicals. They can donate an electron, but then need to gain a replacement from elsewhere before they can continue to function as antioxidants. Such antioxidants are described as being rate limited, as they can only donate electrons at the rate at which they are provided by the cell's metabolism. Vitamin C is different because, being small, safe and water-soluble, it can be supplied to the tissues continuously.

Vitamin C is oxidised in disease

The proportion of reduced to oxidised vitamin C is a measure of tissue health. When a tissue is driven towards oxidation by disease, vitamin C protects it from damage and is oxidised to dehydroascorbate.[176] Conversely, high levels of ascorbate and low levels of its oxidised form, dehydroascorbate, are characteristic of healthy tissue. The ratio of reduced to oxidised ascorbate is central to the action of vitamin C.

In disease states that produce free radicals, ascorbate is oxidised locally to dehydroascorbate. This effect has been shown in many different conditions. Surgery and its associated tissue damage can increase the level of dehydroascorbate relative to ascorbate.[177] Diabetic humans have increased dehydroascorbate levels.[178] Oxygen toxicity increases dehydroascorbate in mice.[179] In rats, dehydroascorbate increases both in diabetic kidneys and in inflammation, but antioxidant supplementation inhibits this effect.[180,181] Inflammation and arthritis are associated with higher levels of dehydroascorbate in mice.[182] The ratio of ascorbate to dehydroascorbate is even used as an indicator of the heath of plant tissue.[183,184]

	Vitamin C and illness		
	Ascorbate	Dehydroascorbate	Tissue
Healthy	High	Low	Reduced
Sick	Low	High	Oxidised

Animals that manufacture their own ascorbate can increase production in response to disease. Sick mice and rats increase their manufacture of ascorbate.[52,53,54] Mice infected with the malaria parasite double their blood levels of ascorbate, while the levels of dehydroascorbate remain constant.[185] These and other animals have a physiological response to illness that increases their ascorbate to dehydroascorbate ratio.

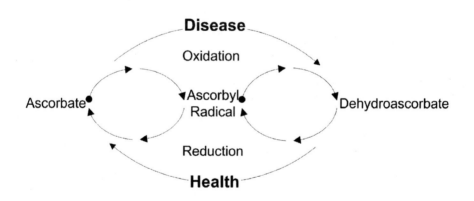

Redox cycle in health and disease

Since tissue damage causes oxidation, vitamin C is converted into the dehydroascorbate form. Therefore, the level of ascorbate (reduced form) is lower and the amount of dehydroascorbate (oxidised form) is increased. Irwin Stone has computed the ratio of ascorbate to dehydroascorbate in several disease states.[437] Stone computed his first table from research published by Bhaduri.[186] As expected, the ratio of ascorbate to dehydroascorbate was found to be low in disease and even lower in severe illness leading to death.

Severity of disease and ratio of ascorbate (AA) to dehydroascorbate (DHA)			
Disease	**Number of subjects**	**Condition**	**AA/DHA**
Normal controls	28	Healthy	14.0
Meningitis	11	Convalescent	2.8
	17	Survived	0.7
	8	Died	0.3
Tetanus	12	Convalescent	5.0
	12	Survived	1.3
	13	Died	0.5
Pneumonia	15	Convalescent	4.0
	19	Survived	1.0
	7	Died	0.4
Typhoid	15	Convalescent	4.5
	19	Survived	1.3
	4	Died	0.4

Stone also calculated the ratio of ascorbate to dehydroascorbate in a range of diseases, based on the work of Chakrabarti.[187] These data are less complete but are consistent with the ratio of ascorbate to dehydroascorbate being an indicator of health or sickness. Stone's results are shown in the following table:

Ratio of ascorbate to dehydroascorbate in disease		
Disease	Number of subjects	AA/DHA
Normal controls	16	14.8
Cholera	21	1.7
Smallpox	16	0.9
Pyogenic meningitis	16	0.7
Tubercular meningitis	16	4.2
Gonorrhoea	16	2.0
Syphilis	16	4.2

This relationship between vitamin C and illness leads us to suggest a number of hypotheses:

- Disease generally depends upon oxidation and involves an excess of free radicals.
- An ample supply of a non-toxic antioxidant, such as ascorbate, can neutralise free radicals and return sick tissue to a reducing redox state.
- A tissue's response to stress, injury or insult is optimal in a reducing redox environment.
- A reducing redox state, induced by ascorbate, modifies signalling by free radicals, modulating the body's immune response by preventing shock and reducing inflammation.

These hypotheses suggest simple experiments that are clearly testable and refutable. Taken together, they imply that if we removed all excess free radicals, the symptoms of a disease would diminish.

The main purpose of ascorbate in fighting disease is as an antioxidant. Vitamin C provides a biologically free supply of electrons to the tissue. Each molecule carries two electrons that can be donated as required to prevent oxidation. In order for this to happen, the ratio of ascorbate to dehydroascorbate must be kept high. This may be achieved by introducing a continuous supply of new ascorbate into the damaged tissue. Incoming vitamin C keeps the tissue in a reduced state, until the unstable dehydroascorbate is hydrolysed (split by combination with water), reduced back to ascorbate, or flows out of the tissue.

Recent measurements of patients with acute pancreatitis confirm these predictions. High doses of vitamin C reduced the number of free radicals, inflammatory chemicals, and the severity of the disease, leading to a shortened the period of illness.[188] This study was unusual, in that it compared a reasonably large 10-gram dose of ascorbate to a one-gram control dose, with both given intravenously.

Bowel tolerance

Through clinical experience, Robert Cathcart discovered that the amount of oral vitamin C tolerated by a sick person, without causing diarrhoea, increases with the intensity of the illness. This observation is easily explained as an evolutionary adaptation, which increases the available ascorbate in times of need. Cathcart's approach to the treatment of disease is to use vitamin C at doses one hundred times the amount normally considered a megadose. The resulting concentration of vitamin C is so large that free radicals are quickly quenched. The tissue is forced into a reduced state and supporting antioxidants, such as vitamin E, are restored to normal functioning. Cathcart warns against giving intermediate doses of vitamin C, suggesting that low megadoses could suppress symptoms while prolonging the duration of the disease, a condition he has described as being "unsick".

Cathcart's well-known table of usual bowel tolerances in disease is reproduced here. Note both the size of the dose and the frequency of administration. This table derives from his extensive clinical experience rather than from controlled clinical trials. Cathcart generalises that tolerance to vitamin C is proportional to the severity of the disease. Generally, this seems to be the case, but it is not always true. Dr Ian Brighthope reports that some very ill people cannot even tolerate half a gram of oral vitamin C, while a number of healthy people can tolerate up to 90 grams without causing observable effects on the bowel.

Cathcart's Table of Usual Bowel Tolerances		
Condition	Grams per day	Doses per day
Normal	4 - 15	4 - 6
Mild cold	30 - 60	6 - 10
Severe cold	60 - 100+	8 - 15
Influenza	100 - 150	8 - 20
ECHO, coxsackievirus	100 - 150	8 - 20
Mononucleosis	150 - 200+	12 - 25
Viral pneumonia	100 - 200+	12 - 25
Hay fever, asthma	15 - 50	4 - 8
Allergy	0.5 - 50	4 - 8
Burn, injury, surgery	25 - 150+	6 - 20
Anxiety, exercise, mild stress	15 - 25	4 - 6
Cancer	15 - 100	4 - 15
Ankylosing spondylitis	15 - 100	4 - 15
Reiter's syndrome	15 - 60	4 - 10
Acute anterior uveitis	30 - 100	4 - 15
Rheumatoid arthritis	15 - 100	4 - 15
Bacterial infections	30 - 200+	10 - 25
Infectious hepatitis	30 - 100	6 - 15
Candidiasis	15 - 200+	6 – 25

We can explain Cathcart's findings, by looking at how the bowel tolerance level operates. When people are sick, their tissues use more vitamin C. This lowers blood levels. When the blood concentration is low, vitamin C is actively transported from the gut. At higher blood levels, the pumps are not effective. During illness, when the blood concentration is low and ascorbate is transported at a higher rate, the concentration in the gut falls. The more severe the illness, the more ascorbate is needed to restore blood levels, and this means a greater dose is absorbed from the gut.

When blood levels have returned to normal, absorption from the gut slows down. If the person continues to take more vitamin C, the concentration in the gut increases. This exerts an osmotic pressure, holding water in the gut, which causes diarrhoea. At this point, we say that the gut tolerance level has been reached. Blood levels are restored just before bowel intolerance is reached. The highest blood levels that can be achieved by oral administration occur around the maximum bowel tolerance level. A reduction in symptoms is often observed as the patient approaches this level.

The capacity of the absorption mechanism is a limiting factor. With severe illness, or in people with a smaller number of ascorbate pumps in their gut, the uptake of vitamin C might not be enough to restore blood levels. In this case, even if the person took sufficient vitamin C to exceed the bowel tolerance level, the illness would continue and blood levels would remain low. It has also been noticed that people who are given intravenous sodium ascorbate have a higher gut tolerance level. The reason for this could be that the intravenous ascorbate counteracts the osmotic pressure exerted by the vitamin C in the gut.

Ascorbate as a drug

The aim of high dose vitamin C treatments is to keep the ratio of ascorbate to dehydroascorbate at a high level, even within sick or inflamed cells. The flow of fresh vitamin C through the tissues means that diseased cells get a large free supply of electrons.

The implications of this idea are considerable. Vitamin C will quench most free radicals it encounters, including other antioxidants that have become oxidised. Furthermore, it will change the redox signalling within the damaged tissue, ensuring that the body provides an appropriate response to the disease. One theoretical problem is that some antioxidants are stronger reducing agents than vitamin C. The electrons used by typical antioxidants come from the metabolism of the cell. When cells supply these electrons, they use energy normally reserved to power the essential chemistry of life. However, large amounts of ascorbate would free these molecules from acting as antioxidants and facilitate normal metabolic functions.[189,190,191,192]

Biological antioxidants are continually engaged in an oxidation-reduction cycle.[193,194,195] This normal recycling uses energy, as it depends on the metabolic pathways of the cell. Damaged or stressed cells may be unable to regenerate sufficient ascorbate or other antioxidants, to cope with the increased demand. From basic chemistry, it is clear that diseased tissue would be returned to a normal reducing state if sufficient ascorbate could be provided and the dehydroascorbate waste product removed.

Clinical experience

In 1975, Cathcart reported that, over a three-year period, he had treated more than 2,000 patients with massive doses of vitamin C.[196] He noted considerable beneficial effects in acute viral disease and suggested that a clinical trial would substantiate his observations. Regrettably, clinical trials of these large doses have not been performed. In 1981, he documented a further 7,000 patients who had been given the treatment, which had markedly altered the expected course of a large number of diseases.[197] Since that time, he has continued treating many thousands of patients, with similar positive results.

Cathcart has observed surprisingly few problems with the massive doses he has tried, stating that the majority of patients have little difficulty with them. This is confirmed by the experience of other physicians giving high dose ascorbate treatment.[2,3,175] Minor complaints such as gas, diarrhoea or stomach acid, reported by healthy people taking large oral doses of vitamin C, are rare in sick patients.

Vitamin C given below bowel tolerance generally has little effect on a disease process, whereas doses that are close to the bowel tolerance level can greatly reduce symptoms. Cathcart describes the effect of vitamin C at these doses as being clinically dramatic, as if a threshold had been reached.[198] Cathcart found that his patients experienced a feeling of "well-being" at high dose levels and considered this an unexpected benefit. These feelings of well-being may indicate that no obvious detrimental side effect is present. Cathcart reports that in severe disease, such as viral pneumonia, the benefit is substantial; he describes a complete cessation of symptoms. Such a powerful effect is difficult to dismiss, either as a placebo response or as self-delusion on the part of the physician.

Importantly, in terms of relating the response to the dose of vitamin C, the symptoms can be turned on or off by adjusting the dose. The sickness and acute symptoms of diseases such as pneumonia were found to return if the vitamin C levels were lowered. This process of switching symptoms on and off with the vitamin C dose is an important observation, since it means that the patients are acting as their own experimental controls. The authors have tried this experiment with the common cold and found that high doses of ascorbate could bring substantial symptomatic relief and a feeling of well-being.

Titration to bowel tolerance levels is fine for people who can be persuaded to take these huge amounts of vitamin C. However, if the disease is more severe or the patient is unable to take large doses orally, then intravenous infusion of ascorbate may be substituted. With intravenous ascorbate, the clinical effect is reported to be more dramatic.[197]

Other physicians and researchers working with massive doses of vitamin C report findings entirely consistent with those of Cathcart. Clinical reports of the beneficial action of these doses are often striking. Patients with severe disease can recover rapidly. An Australian doctor, Archie Kalokerinos, has described seeing children in severe shock, unresponsive to treatment and on the point of death, recovering in a matter of minutes.[575] Either the medical establishment has overlooked an important finding, or multiple independent physicians are each separately discovering a substantial placebo effect that is specific to this vitamin.

Dynamic flow

"If you don't take ascorbic acid with your food you get scurvy, so the medical profession said that if you don't get scurvy you are all right. I think that this is a very grave error."
Albert Szent-Gyorgyi

In this chapter, we introduce the dynamic flow model, which explains how vitamin C acts when large doses are used. An individual with an excess of vitamin C in the diet has a continuous flow through the tissues, which are therefore maintained in a reducing state. It is biologically useful to have a dynamic flow through the body, even though not all the ascorbate is absorbed. During times of stress or infection, ascorbate absorption is increased; the surplus dietary ascorbate then acts as a reservoir upon which the body can draw without delay. When this happens, excretion is also increased, improving the ratio of reduced to oxidised ascorbate in the tissues and helping to restore health.

Dynamic flow model

We have developed a theory of action, the dynamic flow model, which explains the reported clinical results with vitamin C. These ideas build on the pioneering work of Klenner, Stone, Cathcart and Pauling. Using this new model, we are able to clarify both the positive and negative findings that have been reported for vitamin C. The dynamic flow model makes sense of many clinical observations of the effects of ascorbate, over the whole range of intakes. Throughout this book, we demonstrate that the results of experiment are consistent with the model. Indeed, in an extensive search of the literature, we have failed to discover any experimental or clinical studies that do not comply with its predictions.

Dynamic flow restores human physiology to the condition of animals that synthesise their own vitamin C. The majority of animals make ascorbate in either the liver or kidney. The primary role of ascorbate in the body is as an antioxidant; it flows in the bloodstream and diffuses through the tissues providing protection from free radicals. Some cell types have critical requirements for ascorbate and actively accumulate it above the levels found in plasma, using biochemical pumps. When blood levels are low, animals can make more. In addition, ascorbate can be reabsorbed by the kidneys, which prevents levels dropping below a threshold value. Preventing excretion adds a further control mechanism, maintaining minimum levels and conserving energy.

Herbivores can absorb substantial amounts of vitamin C from the food in their guts, reducing the requirement for synthesis. If blood levels increase, manufacture can be suppressed by negative feedback. Normal, healthy animals manufacture relatively high levels of the vitamin, expending considerable energy in the process. The amount made is reportedly equivalent to an intravenous infusion of several grams in man. These animals must therefore either use or excrete this amount.

When animals are sick, they increase both the amount of ascorbate they make and the quantity excreted. By making more, they increase the ratio of reduced ascorbate relative to the oxidised form, dehydroascorbate. As the blood levels rise, high levels of excretion remove dehydroascorbate as well as ascorbate. In the sick animal, both synthesis and excretion increase the levels of ascorbate relative to dehydroascorbate. The homeostatic control of ascorbate during sickness acts to re-establish a reducing internal environment.

Humans and animals that have lost the ability to synthesise their own ascorbate must obtain it from the diet, in the form of vitamin C. In this case, shortage of vitamin C is more critical. Prevention of excretion is essential to avoid scurvy. At low intake levels, humans absorb the vitamin actively from the gut, helping to prevent acute deficiency. When the vitamin is in short supply, the body absorbs a greater proportion of what is available. When there is plenty, some remains in the gut, which acts as a reservoir. A sick human cannot manufacture ascorbate but absorbs more from the gut, if it is available. When ill, people can automatically take in up to 1000 times the levels absorbed by healthy individuals.

If a person is subject to stress or comes in contact with an infective agent, the need for ascorbate to neutralise free radicals is increased. Unless the person has an excess, the available body pool is soon depleted. As the blood levels drop, ascorbate starts being actively pumped from the gut. The amount contained in the digestive system is limited and is soon depleted. An ailing person, who is not aware of the increased need, does not take a supplement and is therefore acutely short of vitamin C.

Ascorbate has a half-life in the blood of about 30 minutes. Paradoxically, this high turnover rate makes ascorbate particularly effective in maintaining a reducing environment in the tissues, provided it can be supplied continuously. Rapid excretion of oxidised vitamin C increases the ratio of ascorbate to dehydroascorbate. Pumps within the kidney reabsorb ascorbate but not dehydroascorbate.[199] Removal of dehydroascorbate from the damaged tissues is important, both for treatment of disease and for animal physiology. The loss of ascorbate in urine is energetically expensive for an animal, but the cost may be balanced by increased survival in the presence of infection and disease.

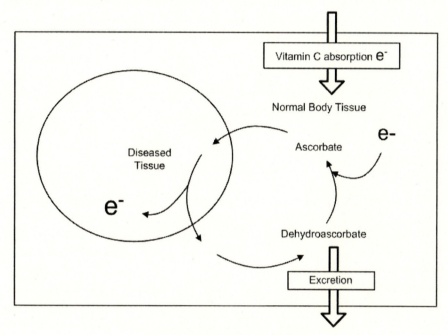

The dynamic flow model

When dynamic flow is achieved, vitamin C passes into diseased tissue as ascorbate, donating electrons (e⁻) and forming dehydroascorbate. The electrons reduce the tissues. A mixture of ascorbate and dehydroascorbate is excreted. The result of this process is a control mechanism that helps prevent tissue oxidation.

In previous chapters, we saw that the body uses small amounts of vitamin C and other antioxidants repeatedly, to neutralise free radicals. The ascorbate molecule cycles round, being continually oxidised and then reduced. This mechanism enables vitamin C to act as a pathway for electrons, linking the cellular metabolism to other antioxidants. The protective oxidation-reduction cycle of antioxidants is limited by the rate at which the cellular metabolism can supply extra electrons. When a person is sick, more free radicals may be produced than a cell can quench, so the cell becomes oxidised and injured. Furthermore, with high levels of oxidants, the cells are damaged and less able to supply the crucial electrons.

Intake levels

In the dynamic flow condition, large amounts of vitamin C are available continuously, to quench free radicals. The dose required for dynamic flow varies with the quantity of free radicals being produced, and is perhaps more than 50% of gut tolerance, spread throughout the day. People opting for higher intakes need to accept the slight risks associated with a larger dose. However, the known risks of, say, a three gram dose of vitamin C are negligible, whereas deficiency is certainly associated with poor health and disease.

Nutritional doses

Acute deficiency means ascorbate intake is extremely low, leading to scurvy. Vitamin C intake of less than about 5mg per day results in severe disease and death.

Sub-clinical deficiency occurs in otherwise healthy individuals when the intake of ascorbate is insufficient to raise levels to the point where white blood cells are saturated and vitamin C is excreted. This level approximates to the RDA, although it would vary with the individual.

Base-level is when the intake of ascorbate is enough to produce blood levels such that vitamin C is consistently found in the urine. This level is necessary to avoid the hypoascorbemia suggested by Irwin Stone.[437]

Dynamic flow is when an excess of ascorbate leads to incomplete absorption and to excretion in the urine. This is necessary to avoid acute vitamin C deficiency in times of stress or disease.

Dynamic flow for disease prevention

When a disease is at an early stage, or is small and localised, a lower concentration of ascorbate is needed than when it is well established. For example, if a person is infected with a common cold, there are relatively few viral particles at the outset. The virus attacks the cells locally, producing an inflammatory response. In these early stages, the volume of tissue involved is small, and the amount of ascorbate required to neutralise the free radicals is correspondingly low. It is easier for the ascorbate molecules to penetrate the diseased tissue than in the later, more widespread stages of the illness. If a person in a state of dynamic flow is exposed to a virus and infection begins to take hold, extra ascorbate is available immediately, from the surplus that is normally not absorbed. This additional vitamin C may often be sufficient to quench the disease, before it has time to become established.

Pharmacological doses

Oral treatment requires ascorbate titrated to bowel tolerance, with large amounts of vitamin C taken by mouth. This appears to be far less effective than intravenous treatment.

Intravenous treatment can be used to administer very high doses of sodium ascorbate for treatment of disease.

Massive doses for treatment of disease

Oral doses, even massive ones, may be ineffective for treatment of disease. Such doses are limited by bowel tolerance and may not be fully absorbed. In this case, intravenous ascorbate offers a useful alternative. Doses given by the intravenous route are limited only by any potentially toxic properties of sodium ascorbate. Fortunately, infusions of several hundred grams seem to be largely free of side effects.[14]

Practical implications of dose levels

The dynamic flow model allows us to make predictions about the effectiveness of different doses of vitamin C. Doses below dynamic flow will not provide adequate protection against susceptibility to disease. We predict that clinical studies using less than about three grams of vitamin C per day will yield variable results. Furthermore, if single doses are given daily, the results will be approximately comparable for all amounts above 500mg, because of the short half-life. Single large doses will only give about four to six hours of protection, as this is the period during which blood levels are raised.

A single megadose will provide only a fraction of the potential benefit of split or slow release doses. Since the action of a drug is generally related to its blood levels, divided doses give greater protection. An immediate consequence of the short half-life of vitamin C is that clinical trials employing a single daily dose schedule underestimate the true benefit. Unfortunately, our review of the literature indicates that the great majority of trials have used single daily doses. This limits the validity of experiments purporting to measure the utility of vitamin C for preventing disease.

Conclusions

The dynamic flow model allows us to take a new look at the claims for ascorbate in disease. It provides an explanation for the positive results. Moreover, it clarifies the variable conclusions of doctors using small or infrequent doses. In addition, it offers a simple reason for why animals that manufacture their own vitamin C also excrete it in their urine. If correct, the dynamic flow model implies that the potential benefits of ascorbate have, so far, been greatly underestimated.

Now that we have an understanding of the role of vitamin C in the body, we are ready to look at its use in specific illnesses. Since the number of diseases that ascorbate is claimed to benefit is large, we will restrict our discussion to the major killers in the industrialised nations.

Heart disease and stroke

"Why think? Why not try the experiment?"
John Hunter

The first disease we will examine in detail is atherosclerosis, the cause of coronary thrombosis and occlusive stroke. Atherosclerosis, heart disease and stroke are arguably the biggest killers in the industrialised nations. Inflammation of the arteries is an important factor in all of them. The role of cholesterol in cardiovascular disease has been overemphasised, however. The evidence for high levels of cholesterol being the fundamental cause is slight.

Searching for a solution

In broad terms, the development of atherosclerosis is comparable to the furring up of domestic water pipes. In arteries, the process involves a complicated assortment of cells and biochemical reactions. Paradoxically, this complexity seems to have narrowed the scope of medical research. Instead of small-scale experiments, designed to identify the fundamental cause of the problem, we find detailed studies of arcane aspects of the biochemistry of lipids (fats or oils).

The preferred research methodology has become large-scale population studies, which are used to identify the relative importance of minor risk factors. While there is obviously a case for detailed scientific information gathering, this process carries its own dangers. The medical establishment have come to believe they are solving an incredibly difficult problem. For this reason, experts in their specialised fields find it hard to accept radical but straightforward new approaches that would invalidate their skills. A solution to the problems of heart disease and stroke could also end the use of techniques around which many doctors have built their careers.

If a simple cure for heart disease and stroke were found, it could also have severe economic implications for the businesses of medicine and pharmaceuticals, along with supposedly health-related food industries, such as margarine producers. Such industries wield a large amount of influence over governments and the medical establishment. Despite these opposing forces, however, an effective solution will eventually move the game forward, potentially leaving professional and economic devastation in its wake. The steady accumulation of detailed research and confusing results, which are ultimately overthrown by a new idea, is a classic process, which has occurred throughout the history of science. Confusion and information gathering are replaced by a new theory that makes sense of the conflicting evidence and explains the facts.

Cholesterol?

The popular and, until recently, conventional explanation of heart disease is that cholesterol builds up on the surface of arterial walls, ultimately blocking the artery and leading to a heart attack or stroke. The root cause of the problem is supposed to be too much cholesterol in the blood, arising from a high fat diet. In practise, few researchers really believe this idea and it is so great an oversimplification as to be misleading. While high blood cholesterol is a risk factor for heart disease, the idea that cholesterol in the diet is the *cause* of atherosclerosis is wrong.

Recently, it has been pointed out that the cholesterol hypothesis is misguided. Critics of the idea cite compelling evidence that people with normal or even low blood cholesterol have heart attacks. Furthermore, there is little evidence linking dietary cholesterol to high blood cholesterol and hence to heart disease.[200,201] The popular model, which involves cholesterol from the diet building up on the arterial wall, has therefore been superseded.

Since few researchers nowadays accept the primitive cholesterol model, simple-minded radicalism is equally misplaced. The involvement of cholesterol in heart disease is now thought to be a result of its oxidation by free radicals. The case against the cholesterol myths is detailed by Dr Uffe Ravnskov,[201] who points out that high blood cholesterol is not closely associated with atherosclerosis during post mortems. Blood cholesterol is also not correlated with the results of coronary angiography (x-ray examination of blood vessels in the heart). The build up of calcium deposits, or calcification, which occurs in older arterial plaques, is not associated with levels of cholesterol or fats in the blood. In addition, high blood cholesterol is not closely related to atherosclerosis in peripheral blood vessels.

"Bad" and "good" cholesterol

The currently popular cholesterol model of atherosclerosis is that excess low-density lipoprotein (LDL) is to blame. According to this hypothesis, LDL, the so-called "bad cholesterol", accumulates in the wall of an artery and undergoes chemical changes, including oxidation, resulting in atherosclerosis. Cholesterol is a fat-soluble steroid, usually called a lipid. As cholesterol is fat-soluble, it does not dissolve readily in water, and the body cannot transport it in the blood plasma. For transportation, it is wrapped with protein molecules into little balls called lipoproteins. LDL is one type of lipoprotein, used to take cholesterol from the liver, where it is manufactured, to the tissues of the body.

The other main form of lipoprotein particle is smaller and denser; this high-density lipoprotein, or HDL, is often called "good cholesterol". It transfers cholesterol back from the tissues to the liver, where it is stored or destroyed. Both LDL and HDL particles are a mixture of proteins and lipid, including cholesterol. The talk of "good" and "bad" cholesterol is both bad science and misleading.

Familial hypercholesterolaemia

The 1985 Nobel Prize for medicine was awarded to Michael Brown and Joseph Goldstein, for their work on the regulation of cholesterol metabolism. They found that fibroblast cells, which produce collagen in the blood vessel wall, have receptors on their surfaces. These receptors allow the uptake of LDL or "bad" cholesterol. Brown and Goldstein discovered that the underlying mechanism in severe familial hypercholesterolaemia (too much cholesterol in the blood) is a lack of functioning LDL-receptors. Familial hypercholesterolaemia is a genetic disease in which blood levels of cholesterol are up to five times higher than normal. In severe cases, where sufferers have a double dose of the gene, they appeared to die from atherosclerosis in their teens or even earlier. People with only a single copy of the gene develop symptoms at around 35 - 55 years of age.

Before we create the impression that the raised blood cholesterol in hypercholesterolaemia causes atherosclerosis and heart attacks, we should clarify the situation. This is what the medical profession considered to be the case, but the facts are not clear. Doctors gained a misleading impression, because the cases of hypercholesterolaemia they observed were often selected because of heart disease. In other words, patients presenting with heart complaints were likely to be identified as having the condition. Others who had the condition but did not have a heart problem were likely to go unnoticed.

The plaques in hypercholesterolaemia can differ from those seen in common heart disease,[202,203] and the cause of death in families with hypercholesterolaemia follows the overall pattern found with heart disease in the normal population.[204] After 1915, the death rate among hypercholesterolaemics increased, reaching a maximum during the 1950s, after which it began to decrease. In the 19th century, the total death rate of people with hypercholesterolaemia was *lower* than that of the general population. If we ignore the possibility that high levels of cholesterol protect against the more common forms of death, this fact refutes the hypothesis that high blood cholesterol causes heart disease. Further, it implies that high cholesterol is not an independent risk factor.

In periods with increased death rates, life expectancy for individual hypercholesterolaemics varied from normal to severely shortened. This large variation in life expectancy suggests that studies of hypercholesterolaemic families, based on selected patients, have overestimated the mortality from the disease. The changes in heart disease death rates over time imply that other environmental factors play a dominant role. Even massively high blood cholesterol does not necessarily cause heart disease, although it is in some way associated with an increased risk. It is interesting that Ginter proposed vitamin C, at high blood levels, as the principal treatment for hypercholesterolaemia.[205]

In normal individuals, cholesterol in the diet inhibits its manufacture in the body. In hypercholesterolaemia, however, there are fewer LDL-receptors on the surface of cells that remove cholesterol, resulting in increased blood levels. This excess LDL cholesterol could accumulate in the wall of arteries, causing atherosclerosis and eventually a heart attack or a stroke. Cells from patients with severe familial hypercholesterolaemia do not have functioning LDL-receptors, so they do not remove the cholesterol from the blood. People who inherit only one gene for the disease have only half the expected number of functioning LDL receptors and thus their blood cholesterol levels are intermediate between people with two disease-carrying genes and normal individuals. In people with hypercholesterolaemia, the cholesterol control mechanisms have failed.

It is easy to see why the idea that LDL cholesterol in the blood causes atherosclerosis was so persuasive. People with hypercholesterolaemia have enormously high levels of blood cholesterol and, before the evidence was fully examined, were believed to die early from complications of atherosclerosis. By analogy, it was assumed that heart disease in normal people also resulted from a defect in the regulation of blood levels of LDL cholesterol. The conclusion was reached that high blood levels of cholesterol causes atherosclerosis and heart disease. This is biology however, and it has no respect for simple logic.

Since high levels of cholesterol occur in arterial plaques, the presence of some form of cholesterol may be necessary for the disease to occur. In severe familial hypercholesterolaemia, people with elevated blood cholesterol can get atherosclerosis abnormally early. In this case, the presence of high blood cholesterol appears to accelerate the progression of the disease. At most, this indicates that high blood cholesterol is in some way associated with the progression of the illness. It does not show that high blood cholesterol is the cause of the disease. If it were that simple, prevention of heart disease could be achieved by controlling blood cholesterol and no one with low cholesterol would ever get the disease, which is not true.

A little thought will supply an alternative to the notion that high blood cholesterol causes atherosclerosis. Fibroblast white blood cells and other cells in the arterial wall manufacture cholesterol. In hypercholesterolaemia, these cells have the same defective receptors as the cells that control cholesterol in the blood. Cells in the arterial wall could continue to manufacture cholesterol, even as it was building up in the wall, because the normal "switch off" mechanism depends on the faulty receptors. In this case, the cholesterol produced locally could build up in the wall, forming plaque independently of the levels in the blood. The cholesterol control mechanism in the blood vessel wall would have failed. High blood levels of cholesterol would be an indicator for the disease but lowering blood cholesterol might not be an effective treatment.

Homeostasis and cholesterol control

Increasing the amount of cholesterol in the diet has little effect on blood levels, since it is manufactured in the liver. Despite its bad press, cholesterol is essential to life and is produced in large amounts by the body, with the remainder coming from the diet. Every cell in the body uses cholesterol in its membranes. Cholesterol deficiency is rare but leads to devastating complications, especially in the central nervous system. It is irrelevant to the body whether the cholesterol comes from the diet or is made internally; the molecule is the same.

In the body, cholesterol is regulated by homeostasis, the name given to the control mechanism that manages most biological substances and processes, keeping them within well-defined limits. The mechanism is a classic negative feedback loop, and it works in the same way that a thermostat controls room temperature. If too much cholesterol is supplied in the diet, the liver can reduce its synthesis of new cholesterol or, alternatively, increase cholesterol breakdown. Both these mechanisms regulate the amount of cholesterol in the body. If the body's cholesterol levels are too high, then the control mechanism is not working properly.

Reducing the intake of dietary cholesterol may reduce blood levels, but greater production by the liver can compensate. Because of homeostasis, cutting a person's cholesterol intake would stimulate the body to manufacture more in the liver, to make up for the reduction.

Dietary cholesterol and heart disease

The enormous amount of effort expended on the hypothesis that dietary cholesterol is the main cause of heart disease has effectively refuted the idea. Together with instructing patients to reduce their intake of cholesterol, the conventional emphasis is now on the replacement of saturated fats with polyunsaturated fats in the diet. This is also scientifically controversial.

In medicine, there is often an implicit assumption that high blood cholesterol levels lead directly to arterial plaque formation: this can lead to confusion. For example, a daily one-gram dose of vitamin C was shown to increase blood cholesterol in patients with atherosclerosis. Initially, this rise was supposed to be wholly due to removal of cholesterol from the arterial walls and plaques,[206] and the suggestion was made that, in this case, the risk of heart attack would be reduced as blood cholesterol rose. Later calculations suggested that this increased blood cholesterol might not all come from the arterial wall.[207] If the increased cholesterol came from some other source, it was argued, it might lead to increased risk. Of course, this argument is based on the covert assumption that high blood cholesterol causes atherosclerosis. Initial studies in healthy young patients suggested that vitamin C lowered blood cholesterol, although later studies were unable to confirm this result.[208] Such studies as these throw little light on the effects of either vitamin C or cholesterol on heart disease.

Atherosclerosis

Plaques consist of fatty tissues that accumulate within the arterial wall; they are a major sign of atherosclerosis. As a plaque forms, cholesterol is deposited within it. Unlike the furring up of water pipes, plaque formation is an active process. Initially, the artery wall often responds by thickening locally and expanding in diameter, this keeps the blood flowing as the plaque grows. The thickening of the wall of an artery supplying the heart can prevent it from expanding with increased blood pressure when a person exercises; the result is angina pectoris, a feeling of tightness and pressure in the chest. In other arteries, it can lead to tightening and pain in the legs, known as intermittent claudication.

Eventually, the plaque begins to obstruct the vessel and reduce blood flow; this constriction is called stenosis.

In some cases, the plaque can continue to expand until it totally blocks the artery, preventing the flow of blood. Only about 15% of heart attacks result from direct blockage by a growing plaque. Most are caused when the plaque ruptures, producing a crack or fissure in the internal wall of the artery. Blood then comes into direct contact with the damaged tissue and forms a clot, just as it would at the site of any other injury. The production of such a clot can have the unfortunate result of completely preventing blood flow. Alternatively, it can break off and block a vessel elsewhere.

Sugar: glucose and glycation

The popular misconception of a direct causative link between dietary cholesterol and heart disease has encouraged people to eliminate cholesterol from their diet. However, there is more reason to reduce sugar in the diet than cholesterol. Dr John Yudkin has publicised the involvement of sugar in heart disease.[209] He described the relationship between the intake of sugar and rates of coronary heart disease. Atherosclerosis, heart disease and stroke are well-known complications for diabetics with increased blood glucose levels.

In arterial plaques, molecules in the blood vessel wall are cross-linked by glucose, a process known as glycation. Glycation is a factor involved in high blood pressure and the aging of arteries. Glucose acts as a bridge between collagen and other molecules in blood vessel walls, interfering with their normal elastic function and making them stiffer. The reduced elasticity of the vessel wall means that normal expansion of the blood vessels requires higher blood pressures. Glucose is thus a significant factor in arteriosclerosis, the aging and hardening of arteries. As the arteries stiffen, stress on the arterial walls increases, especially in areas of high curvature and fast blood flow.

Plaque development

In the inflamed arterial wall, LDL cholesterol undergoes two denaturing processes: oxidation and cross-linking of its proteins by glucose. The denatured LDL stimulates the inflammation process further and a fatty streak is formed. This streak consists largely of oxidised LDL, inflammatory molecules and white blood cells. Given time, the streak becomes a plaque, composed of a fatty core covered by a fibrous cap. Smooth muscle cells migrate to the surface and multiply, laying down collagen fibres to cover the plaque with a cap. The building of the fibrous cap enlarges the plaque, increasing the likelihood of artery blockage. By strengthening the plaque, the cap also minimises the chance of rupture and clotting, and prevents heart attacks and strokes in the short term, at least.

White blood cells, especially macrophages, initiate further inflammation. As the plaque grows, the macrophages take up so much fat that, under the microscope, they appear foamy. For this reason, they are called foam cells. With time, the increased inflammation can weaken the fibrous cap, leading to rupture and clot formation.[210] As well as blocking the blood vessel in which it is located, a clot can break off and travel through the blood, until it reaches and obstructs a small vessel that it cannot pass through. If the clot is big enough and is in one of the arteries supplying the heart, a coronary thrombosis can result. A clot that breaks off and lodges in a vessel supplying the brain produces a stroke.

Plaque location

Plaques occur more frequently in places that are stressed, often near the heart where blood vessels stretch and bend. High blood pressure and the pulsating flow tend to flex the blood vessels in these areas. The flow of blood around an obstruction also causes mechanical stress to the cells lining the artery. Oxidation associated with this mechanical stress can cause white blood cells to adhere to the arterial wall, an early stage of plaque formation.[64,211] These mechanical and other stresses on the artery produce inflammation and stimulate plaque formation. In some cases, a blood vessel can even rupture because of the local damage.

Plaques as inflammation

Atherosclerosis is now recognised as an inflammatory disease.[212,213,214] Arterial plaques are active structures and may flare up or shrink back, depending on the local inflammatory state. This revised model of heart disease as inflammation helps to explain why LDL cholesterol is a risk factor. The problems start when LDL particles collect in the internal lining, or intima, of the artery. This lining is formed by a thin layer of cells, together with some muscle cells, in connective tissue. Lipoprotein particles are able to pass into and out of the arterial lining. Sometimes, however, the LDL can begin to accumulate in the arterial wall. Surprisingly, this accumulation is not a problem in itself, as LDL is a normal part of the body and does not cause inflammation. However, when it is oxidised or cross-linked, it changes its molecular shape and takes a form that the immune system recognises as abnormal, triggering inflammation.

Derived from solid scientific evidence, the inflammatory model of atherosclerosis, heart disease and occlusive stroke is consistent with the known facts. The cellular processes involved in the formation of plaques are similar to those in other chronic inflammatory diseases that involve fibrosis (the replacement of healthy tissue by scar tissue).[215] Recently discovered markers of inflammation, such as C-reactive protein, are being proposed as indicators of heart attack risk in clinical practice.[216,217] The model of heart disease as inflammation is becoming mainstream.

Infection and heart disease

The role of infection in heart disease, which might once have been considered a crackpot idea, is being taken seriously,[218] and is supported by reports that both bacteria and viruses have been found in plaque tissue.[219] Now that researchers are actively looking for the presence of infection, the evidence for its involvement is increasing. There is even cautious consideration of the treatment of heart disease with antibiotics.[220]

If heart disease can result from infection, the discussion in later chapters on vitamin C and infectious diseases becomes directly relevant. For example, there are at least three ways to explain the recognised link between gum disease and atherosclerosis.[221] The first is that gingivitis-causing microbes occasionally travel in the bloodstream and trigger inflammation of the arteries. The second is that a low-level infection of the gums secretes chemicals into the blood stream that promote plaque formation. The third mechanism, which seems closer to being causative, is that sugar and low antioxidant levels in the diet produce both gum and heart disease.

Anti-inflammatory drugs

Aspirin is a well-known anti-inflammatory drug, thought to prevent heart attacks by either "thinning the blood" or inhibiting clot formation. In inflammatory conditions such as arthritis, aspirin is generally taken at higher doses than those used to prevent heart attacks. It is of interest that the statin drugs, which are thought to prevent heart attacks by reducing the levels of cholesterol in the blood, also have anti-inflammatory and other actions.[222,223,224] Indeed, lowering blood lipids and cholesterol reduces plaque inflammation in rabbits and other animals,[225] providing an alternative explanation for the proposed benefits. Drugs and interventions often have more than one biological effect and it is possible that these "cholesterol lowering" drugs actually prevent heart attacks by acting to reduce inflammation, stabilise arterial plaques or reduce clotting.

Oxidation and free radicals

A fundamental cause of inflammation in atherosclerosis is the action of free radicals.[226] In the blood vessel wall, free radicals released by white blood cells oxidize lipids and other molecules. This oxidation is a primary factor in the formation of plaques.[227] Both lipids and proteins are oxidised because of higher levels of free radicals in the diseased arterial wall.[228] Free radicals stimulate the increased cellular growth that leads to plaque formation.

Vitamin C and other antioxidants confer numerous benefits on heart and blood vessels. Vitamin C is required for the production of collagen, which helps to strengthen the blood vessel walls. The addition of antioxidants to the diet can prevent lipid oxidation[227,228] and vitamin C stops LDL cholesterol from being oxidised in blood plasma,[229,230] both directly and by regenerating vitamin E. Vitamin C is claimed to be the outstanding antioxidant in blood.[231] Antioxidants lessen the injuries caused by reduced oxygen supply and reperfusion damage in heart disease and stroke.[36,37] As described previously, reperfusion damage happens when the blood supply returns to a tissue after a period of oxygen starvation. An understanding of the processes of inflammation leads directly to the expectation that vitamin C and other dietary antioxidants are beneficial in heart disease.[232,233,234]

The oxidation theory of atherosclerosis

An extension of the inflammation model of atherosclerosis is the oxidation theory. The oxidation theory does not invalidate any of the claims for the involvement of inflammation or infection; it simply suggests that plaques arise principally from oxidation induced by free radicals. Scientists believe oxidation to be an important factor in the development of atherosclerosis.[235] Blood rushing past obstructions or bends can cause oxidative stress to the cells lining the artery.[236] High blood levels of the amino acid homocysteine also damage arterial walls,[237] by producing free radicals in the arterial wall, although the evidence is not complete.[238] Homocysteine is produced from the amino acid methionine, when the diet is deficient in vitamin B6, folic acid and other nutrients.

In plaques, oxidised protein and LDL cholesterol sticks to the blood vessel wall, leading to free radical damage.[239] Oxidised lipids induce inflammation,[243] and cholesterol itself can be oxidised, producing a chemical that is toxic to cells in the arterial wall.[240] Oxidised LDL cholesterol may cause blood coagulation, and promote calcification in advanced plaques.[241] Macrophage white blood cells cause oxidation that disrupts the plaque, leading to clot formation and heart attack.[242] While macrophages take up oxidised LDL cholesterol,[243] this can cause harm and cell death.[243,244] Dying macrophages cause yet more oxidative stress in the tissue and contribute to the deposition of sludge in advanced plaques.[245]

Preventing oxidation of "bad" cholesterol

It is generally accepted that plaque formation could be modified by the addition of suitable antioxidants. However, many dietary antioxidants may be unable to penetrate the plaque at sufficient concentration to have an effect. Within the plaque, LDL cholesterol has limited resistance to oxidation,[246,247] although it contains small amounts of antioxidants, such as vitamin E and coenzyme Q10.[36] Furthermore, HDL, or "good" cholesterol, can also be oxidised.[248,249]

Isolated LDL cholesterol can be oxidised using copper ions, Cu^{2+}. When this happens, there is a lag period, during which the antioxidants in LDL are oxidised before the other constituents. Adding vitamin E can significantly increase the resistance of LDL cholesterol to oxidation. However, with prolonged exposure to free radicals, the LDL cholesterol is oxidised, so the protective effect is limited. When vitamin E in LDL cholesterol is oxidised, the free radical formed can react with lipids causing further oxidation damage.[250] In the absence of ascorbate to recycle it, vitamin E may cease to be protective and could promote free radical damage.

Vitamin C can prevent the oxidation of LDL cholesterol directly,[251,252,253,254,255] and can also help other antioxidants stay effective.[64,230,256,257,231,258] High levels of ascorbate can immerse the cholesterol lipoproteins in a reducing liquid. However, despite ample biochemical evidence for a protective effect of dietary antioxidants, it has generally been claimed that they provide limited benefit.[212] Presently, we will show that this claim results from a misinterpretation of the science.

Heart disease or scurvy?

"Man will occasionally stumble over the truth, but usually manages to pick himself up, walk over or around it, and carry on." Winston S Churchill

More than sixty years ago, a Canadian pathologist called Paterson put forward the hypothesis that shortage of vitamin C is a causative factor in atherosclerosis. Since then, the hypothesis has failed to enter the medical mainstream. Several independent scientists, including Linus Pauling, concluded that heart disease is a form of scurvy. In 1976, Turley and colleagues stated that, "Strong clinical and experimental evidence suggests that chronic latent vitamin C deficiency leads to hypercholesterolaemia and the accumulation of cholesterol in certain tissues".[259] In other words, they believed that shortage of vitamin C had already been shown to cause atherosclerosis.

Similar theories explaining heart disease as scurvy have been proposed more recently by Price,[260] and by Clemetson.[261] Price's group suggested that latent scurvy causes a reversible atherosclerosis, similar to the clinical form of the disease in man. They also proposed that an excess of glucose, which is structurally similar to ascorbate and can inhibit its uptake into cells, could induce a form of local scurvy. Clemetson suggested that shortage of vitamin C disturbs cholesterol metabolism and raises levels of histamine, damaging the lining of the arteries and predisposing the person to atherosclerosis.

Do animals get heart attacks?

Atherosclerosis is common in humans but rare in domesticated mammals. People are told to watch the amount of fat in their diet, for fear of getting heart disease. They do not get the same advice for their pets, because animals do not generally get occlusive heart disease. Such disease is not an important consideration for vets treating dogs, cats, rabbits and gerbils, for example. It does not matter whether the animal is carnivorous or herbivorous; the disease is uncommon in both.

You might think that doctors would have been stimulated by the absence of atherosclerosis in animals, to find the defensive mechanism involved. One obvious biochemical difference is that man has lost the ability to manufacture ascorbate, whereas the animals that do not get heart disease make it internally in large amounts. In guinea pigs, which are unable to synthesise ascorbate, the situation is quite different. Unless their diet includes sufficient vitamin C, they do suffer clogging of the arteries. These ascorbate deficient animals form plaques, despite not ingesting large amounts of cholesterol.

Dogs and cats have a limited ability to produce their own vitamin C. Weight for weight, dogs produce vitamin C equivalent to about 2.5 grams in humans. In an x-ray review of 3022 dogs and 671 cats, only 19 dogs showed coronary mineralisation, a sign of developed plaques.[262] No cat had any mineralisation. There are indications that pet dogs may sometimes be vitamin C deficient. They suffer from illnesses, such as hip dysplasia, that are correctable with supplemental vitamin C.[437] The occasional occurrence of atherosclerosis in dogs could be a result of diabetes, in which high blood sugar would predispose them to the condition. An alternative explanation involves infection, as chlamydia has been found in both dog and human arterial plaques.[263] It is possible to produce a degree of atherosclerosis in the domestic cat, by feeding it a diet abnormally high in fat for several months.[264] Atherosclerosis is, however, rare in dogs and even less frequent in cats. Goats produce much more vitamin C than cats or dogs,[265] and are believed to increase production when stressed or infected. The increase is equivalent to going from 13 to 100 grams in a human.[422]

Mathias Rath has suggested that domestic animals *never* get atherosclerosis. This is overstating the case; domestic animals sometimes get the disease and the frequency can be increased in some cases by adding cholesterol to the diet. Pigs can develop a level of atherosclerosis when fed a high fat diet but, like monkeys, their plaques do not usually develop local blood vessels, or rupture, like those in man.[266] Pigs are used as an animal model for atherosclerosis research. As we might expect, diabetic pigs get atherosclerosis.[267] Birds, especially old parrots and cockatoos, can also suffer from atherosclerosis.[268] Baboons get a form of atherosclerosis in the wild but do not make good animal models of the disease.[269] Wild monkeys also can get a degree of atherosclerosis,[270] which can be worsened, in captivity, by feeding them a high cholesterol diet.[271] The natural form of atherosclerosis observed in mammals other than primates differs from that in man, in having less fatty deposits.[272]

Rath has claimed that bears, in particular, do not get heart disease. He suggests that high levels of vitamin C are protective, when blood fat levels increase during hibernation. Hibernating animals can modify their vitamin C levels during hibernation.[273] However, Bensky, an Asiatic Bear (Selenarctos thibetanus) from Little Rock Zoo in Arkansas, died suddenly on the 8th May 2002 from ischemic heart disease due to severe coronary atherosclerosis. Bensky, aged 23 years, was being gradually introduced to his new mate, Samone. He had been with his previous mate, Sugar Bear, from 1975 until her death in 1998. This report suggests that bears can die from coronary heart disease, although the incidence is probably low.

Most animals are resistant to the development of atherosclerosis and heart disease. Where it does occur, it is likely that atherosclerosis in animals is linked to a limited internal production or utilisation of vitamin C and other antioxidants, secondary to diabetes or some other ailment. A study on the effect of a high cholesterol diet has shown that it does not affect antioxidant status in the rat, which is resistant to atherosclerosis, but markedly affects antioxidant status in the quail, which is susceptible to plaque formation.[274] The existence of the disease in occasional animals, or its induction by feeding them abnormal amounts of cholesterol or fat, does not alter the facts. The evidence suggests that animals with high levels of vitamin C are resistant to atherosclerosis.

Fragile plaques

In 1940, pathologist J.C. Paterson from Ottawa was studying atherosclerosis and capillary rupture leading to strokes.[275] He noted that stroke and coronary thrombosis were associated with lumps in artery walls, called plaques. Blood clots could break off these plaques and block blood vessels in the heart and brain. His histological studies led him to believe that the blood clots were a result of damage to the capillaries in the region of an atherosclerotic plaque. This account is similar to modern theories.

The following year, Paterson carried out microscopic examinations of plaques, which led him to suggest that high blood pressure stresses the blood vessels and damages capillaries.[276] He added that the fragility of capillaries in plaques might be related to vitamin C deficiency. He measured ascorbate levels in patients with coronary occlusion, and found that 81% had levels of less than 0.5mg per 100cc of plasma, which was low. This incidence of vitamin C deficiency was not matched by any other illness. Paterson knew that vitamin C deficiency could result in capillaries becoming fragile and easily damaged, as happens with scurvy. He suggested that inadequate blood levels of vitamin C might be a cause of coronary thrombosis. He concluded, "There is sufficient evidence to warrant the recommendation that patients with coronary artery disease be assured an adequate vitamin C intake either by a proper diet or by the exhibition of ascorbic acid, an innocuous drug".

If, as Paterson suggested, atherosclerosis were caused by shortage of ascorbate, then people with scurvy would be particularly at risk. An interesting story on the internet reports an early study of vitamin C deprivation, and claims that half the subjects had heart attacks. This study was conducted in Sheffield, England by a team led by the Nobel Prize winner, Sir Hans Krebs.[277] The subjects were conscientious objectors, who volunteered during the war years, 1943-1946. In 1996, this study was incorrectly included in a discussion of medical ethics and the Nuremberg doctor trials, because of the supposed death of one of the subjects.[278] When we tracked down the details, we found that one of the original researchers, Dr John Pemberton, was still active at 91 years of age. Pemberton told us that 10 volunteers were deprived of vitamin C and all developed scurvy, over a period of six to seven months. None of the subjects died during the clinical trial. However, two showed symptoms of heart attack and recovered when vitamin C was given.

Current research also supports Paterson's ideas. Vitamin C strengthens plaques, making them less liable to rupture.[279,280] Lack of ascorbate weakens the plaque and this, combined with the stress from blood flow, increases the likelihood of clot formation.[281] In 2002, Nakata carried out work using a strain of mouse that could manufacture neither vitamin C, nor a constituent of very low-density lipoprotein (LDL) cholesterol.[282] The plaques in these deficient mice contain less collagen and are more likely to rupture than those normally found in atherosclerosis. Vitamin C supplementation stabilised the plaques and made them less liable to clot and block blood vessels, as Patterson had suggested. However, supplementation did not alter the number of plaques. While this result indicates that vitamin C could lower the frequency of heart attacks, these mice are an unusual strain and may be unsuitable as a model for humans.

Paterson believed that vitamin C deficiency contributed to the development of heart disease and strokes. He found that vitamin C levels were lower in patients with coronary occlusion. He concluded that vitamin C research was still in its infancy and clinical evidence would need to accumulate in order to verify his findings. Given this promising start, subsequent progress has been disappointing. However, another Canadian, Dr C.G. Willis from Montreal, followed up Paterson's observations, with rather striking results.

Stressed arteries form plaque

In 1953, Dr Willis carried out a study of the arterial wall in atherosclerosis.[283] His observations supported Paterson's suggestion that atherosclerosis was caused by vitamin C deficiency. He pointed out that, in patients with scurvy, fat was deposited in the arterial wall and noted that lesions with fragile capillaries occur at sites of mechanical stress.

If these ideas of Paterson and Willis were correct, then we would predict that vitamin C deficiency would lead to localised damage of blood vessels. This was recently confirmed by Dr. Louis J. Ignarro, who gained the 1998 Nobel Prize in medicine for work on nitric oxide signalling in the cardiovascular system. Ignarro found that dietary supplements of antioxidant vitamins and L-arginine, an amino acid found in the normal diet, could lower the risk of heart disease in mice.[284] In 2003, he suggested that atherosclerotic plaques act like trash caught in a river bend, impeding the flow. He proposed that treatment with antioxidants and L-arginine could prevent blood vessel inflammation and subsequent damage. While his experiments were on mice with a genetic defect in cholesterol metabolism, he believed his experiments implied the same outcome in patients with atherosclerosis and heart disease.[285]

Atherosclerosis in scurvy: animal studies

Willis predicted that if he deprived animals of vitamin C they would get atherosclerosis. He selected the guinea pig for his experiments[286] and fed 145 animals a high dose of cholesterol, together with varying amounts of vitamin C.[283] Guinea pigs are herbivores and do not normally consume cholesterol in their diet. Unlike humans, these animals do not have a good control mechanism for cholesterol. In the guinea pig, high consumption leads to cholesterol being deposited throughout the body.

As predicted, the animals given insufficient vitamin C rapidly developed a form of atherosclerosis, reported to be indistinguishable from the human form. High levels of cholesterol also led to plaques in all animals that were short of vitamin C. When dietary cholesterol levels were high, vitamin C supplementation or injection reduced the incidence of plaques. As expected, injected vitamin C was more effective and prevented plaque formation in about half the animals. The plaques found in animals fed both vitamin C and cholesterol were less severe than those in the ascorbate deficient animals.

It is now established that guinea pigs, one of the few animals known to require dietary vitamin C, get atherosclerosis if they are fed a deficient diet.[287] Linus Pauling and Mathias Rath have replicated and confirmed this early work. In a pilot study, they experimented with three guinea pigs and found that they could induce atherosclerosis by reducing the amount of vitamin C in the diet.[288] In a further experiment on 22 animals, they confirmed that atherosclerosis occurred in guinea pigs fed low ascorbate diets. Other studies have shown that inadequate intake of vitamin C in the guinea pig leads to high blood cholesterol,[289,290] high blood lipids,[291] atherosclerosis,[292] and increased oxidative stress.[293] It has also been confirmed that a high cholesterol diet with low ascorbate produces atherosclerosis in guinea pigs.[294] While a link between dietary cholesterol and heart disease is not supported by the evidence in humans,[295] the established relationship between low vitamin C and atherosclerosis in guinea pigs may reflect a cause of the disease.[286]

Willis' results suggest that vitamin C would stop plaques forming in other experimental animals. It is possible to produce atherosclerosis experimentally in rabbits, by feeding them an unnaturally high cholesterol diet. While rabbits are able to produce their own vitamin C, they may not produce all they need.[422,296] Wild rabbits live on a vegetarian diet, which supplies additional vitamin C, but chow pellets replace this in most experiments. Since some commercial chow does not include vitamin C, experimental rabbits may have insufficient for dealing with the stress of a high cholesterol diet.

As predicted, vitamin C inhibits the formation of plaques in rabbits. Finmore fed cholesterol to rabbits and found that vitamin C reduced the amount of cholesterol deposited in the aorta.[297] He claimed that vitamin C had a specific action on cholesterol in arterial walls. In a later study, he found that vitamin C did not diminish the extent of existing plaques in rabbits.[298] These results on the protection of rabbits from atherosclerotic plaques by vitamin C are confirmed by other research[299,300,301,302] but not by all studies.[303,304] Other antioxidants can also slow down plaque formation in the rabbit.[305,306] In one report, vitamin E and selenium were found to prevent plaque formation but vitamin C was not.[307] Despite this finding, the inhibition of atherosclerosis by vitamin C in rabbits is generally accepted.[308,309] Rabbits can develop atherosclerosis on an unusually fat laden diet, but additional vitamin C may prevent plaque formation, even in this animal, which manufactures its own.

We can also predict that if an animal that normally manufactured ascorbate lost this ability, it would get plaques. Consider the example of mice, which normally produce their own vitamin C and are resistant to atherosclerosis. A particular laboratory strain of mouse cannot produce ascorbate, as it has a mutated form of gulonolactone oxidase - the same enzyme that is missing in humans. If not supplemented with vitamin C, this mouse suffers damage to the aorta.[310] The injury does not have the same characteristics as human plaque, but evolutionary changes may have compensated for the lack of the vitamin in humans. Another strain of mouse, with an inability to process cholesterol effectively, is subject to atherosclerosis, which is inhibited by supplementation with vitamin C and other antioxidants.[311] These mice have very high cholesterol levels, like hyperlipidaemic humans.

Reversing atherosclerosis in animals

Once a plaque has grown, it may be difficult to remove. In 1957, Willis tried to demonstrate the reversal of atherosclerosis in the guinea pig.[312] Willis fed an ascorbate deficient diet to 77 animals. After periods ranging from 21 to 30 days, he gave vitamin C to 50 of the animals and examined the remaining 27 for signs of atherosclerosis. He fed the same diet, but with vitamin C added, to a control group of 12 animals. As expected from his earlier results, no atherosclerosis was found in the vitamin C supplemented controls, whereas 16 of the 27 ascorbate-deficient animals showed signs of atherosclerosis. He found that 34 of the 50 animals that had received vitamin C for varying periods, following the deficient diet, had developed atherosclerosis. In his discussion, Willis suggests that the less developed plaques, or "early lesions", were quickly reabsorbed following vitamin C treatment. Large advanced plaques were considerably more resistant to reversal.

Willis claimed that when he fed vitamin C to his guinea pigs, small plaques were healed quickly. After only two days of treatment, he observed that plaques early in their development appeared to be in regression. In more advanced plaques, the large lipid centre was broken up into islands of fat but, after nine days of treatment, he noted little further change. Willis' results suggest that vitamin C can prevent atherosclerosis in guinea pigs but will only partially remove it, leaving scar tissue behind.

Clinical reversal of atherosclerosis in humans

If vitamin C were to be a useful treatment, it would need to be effective in people as well as animals. In 1954, Willis began a study of the development of plaques in the femoral (thigh) and popliteal (back of the knee) arteries, using serial x-ray angiography (x-ray examination of blood vessels).[313] This technique allowed him to obtain repeated images of the same plaques, to see if they had changed over time. He noted that if these arteries were atherosclerotic, then experience suggested the disease was spread throughout the patient's vascular system. His patients had generalised atherosclerosis and varied in age from 55 to 77 years, with an average of 64 years. The treatment group of 10 patients was given half a gram of vitamin C orally three times a day, while the control group of six patients was not. Nowadays, 1.5 grams of vitamin C is considered a nutritional rather than therapeutic dose, but giving three doses each day would have raised the average blood levels.

After their initial x-ray assessment, Willis gave his patients a repeated arteriograph, following a period of between two and six months. In patients who were treated with vitamin C, Willis noted that plaques, which had previously been growing bigger, were becoming smaller. The patients' symptoms also improved. In the vitamin C group, six patients improved and their plaques reduced in size, three cases showed plaque growth and one patient had plaques of constant size. The untreated controls showed no change in symptoms and the plaques grew larger in two of these patients. None of the unsupplemented controls showed any improvement in their disease.

Willis' clinical study is open to criticism of potential bias in patient selection, the placebo effect, experimenter prejudice and for the qualitative nature of his measurements. He felt his results were encouraging, but pointed out that the series was small and that conclusions must await studies carried out for a longer time and with more cases.

More recently, x-ray scanning measurements support Willis' conclusions. Mathias Rath and Aleksandra Niedzwiecki measured arterial deposits repeatedly, in 55 subjects, using an ultra fast x-ray CT body scanner. They studied a therapy based on high dose vitamin C and the amino acid lysine, and found that it greatly reduced the growth of deposits in the first six months of treatment.[314] In the following six months, in patients with early atherosclerosis, growth of arterial plaque ceased. Their paper presents x-ray images suggesting reversal of plaque deposits. However, x-ray measurements of this form are difficult to perform and are prone to error. Taking measurements from x-ray CT scans depends on many factors, such as whether the imaged slice through the heart is at exactly the same level each time. Like those of Willis, these measurements may be subject to investigator bias.

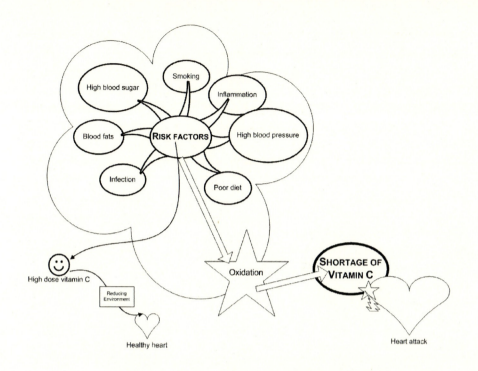

Vitamin C, risk factors and heart disease

Prevention of plaque growth

Scientists have confirmed that ascorbate can prevent growth of arterial plaques in humans. Vitamin C has been used following surgery, to prevent arterial blocking. In a preliminary study to assess the possibility of using vitamin C to prevent arterial blockage following a surgical procedure, Tomoda found the incidence of re-blockage was significantly reduced in 50 patients receiving 500mg each day.[315] Fang studied 40 patients, given either 500mg of vitamin C and 400IU of vitamin E, or a placebo, twice daily for one year, following heart transplants.[316] He evaluated the effect by measuring the area of the plaque using intra-vascular ultrasound. Over the year of treatment, plaque size in the placebo group increased by 8% but did not change in the treatment group. These data provide independent supporting evidence that the antioxidant vitamins C and E can retard the progression of atherosclerosis.

Low dose studies

The scientific literature on low dose vitamin C supplementation and heart disease shows considerable variation. This variation is predicted by the dynamic flow model, given the size and timing of the dose and the initial condition of the subjects' arteries. Since Willis performed his experiments, most studies of vitamin C and heart disease have been indirect and have used low doses. Despite this, there is substantial evidence that low doses of vitamin C can inhibit arterial plaque formation,[317,318,319,320,321,322,323] although it is not conclusive.[324,325,326]

A randomised-control, three-year clinical study in Kuopio, Finland showed that vitamins C and E slowed down the progression of arterial plaque growth in men.[327] This study used two daily 250mg doses of slow release vitamin C and 136 IU of vitamin E. While this is a low dose of vitamin C, the use of slow release tablets helps to sustain blood levels. The combination of vitamins was effective, but the effect of either vitamin C or vitamin E alone was not significant. This suggests a synergistic benefit, as these antioxidants are known to interact. A follow up report three years later indicated that the beneficial effect had continued for six years.[328] Another study sampled blood levels of vitamin C in 468 male and female subjects, aged 66-75 years, living in Sheffield, Yorkshire. The researchers measured the thickness of the arterial wall using ultrasound. The results showed that in men, higher levels of vitamin C were associated with thinner carotid artery walls.[329] The effect was not replicated in the women in the study.

In 1996, Lynch found evidence that vitamin C can slow the development of experimental atherosclerosis in animals.[330] He argued that the current evidence from epidemiological studies on vitamin C in the prevention of cardiovascular disease is not conclusive. Some studies show a strong correlation between increased vitamin C intake and reduction of cardiovascular events, while others do not. That same year, Ness also reviewed the literature on the use of vitamin C in atherosclerosis and heart disease.[331] The review included studies published between 1966 and 1996. Ness concluded that the limited evidence available was consistent with vitamin C protecting against stroke, but the evidence for coronary heart disease was less consistent. He suggested that the relative lack of a link to coronary heart disease could be explained by too low a dose of the vitamin being used.

Vitamin C levels in plaques and arteries

In 1955, Willis and a scientific colleague decided to measure the ascorbate content of human arteries. He concluded that ascorbic acid depletion is common in human arteries, but suggested this deficiency could be remedied by vitamin C therapy. A study in 1995 by Suarna noted that Willis' work on measurement and implications of vitamin C in arteries had not been followed up, in the intervening decades of research into heart disease.[332] This later study indicated that human plaque might contain more vitamins C and E than is found in healthy arteries. In plaques, however both vitamin E and coenzyme Q10 were in the oxidised form.[333] The antioxidants within the plaque were insufficient to force the tissue into a reducing state. We can be confident that plaques are in an oxidised state, which could be reversed with sufficient vitamin C.[334,335,336,235,337]

What happened to this research?

The medical establishment has failed in its obligation to follow up these findings effectively. The animal studies are now arguably definitive; ascorbate prevents heart disease in animals. Willis' animal studies have been replicated and the relationship between antioxidants and plaques in animals is established. In humans, the experiments have not been carried out.

When Paterson and Willis were doing their research into vitamin C and heart disease, they presumably believed that it would be replicated with solid clinical trials. In the intervening decades, the medical profession spent millions on cholesterol research, despite good initial evidence that the cause of atherosclerosis was not excess cholesterol, but deficiency of vitamin C. Unfortunately, half a century later, we are still waiting for simple, inexpensive experiments to be performed. The medical profession has failed to either validate or refute this treatment for the biggest killer in the industrialised nations.

The effects of vitamin C on atherosclerosis have been studied at several levels. These include test tube chemistry, physiological experimentation and clinical trials. If the hypothesis that heart disease is a form of scurvy is true, why have these previous studies failed to provide the crucial data? Is it possible that all these highly trained investigators could have missed such a large effect? Surprisingly, the answer is yes. The reason is that almost all the thousands of medical papers published in this area have used doses too small to be effective.

It would be relatively simple to test this hypothesis. We would need to measure the extent of blood vessel blockage, both before and after treatment with high dose vitamin C or placebo, in a small number of people with heart disease. Measurement of blood flow and wall thickness in vessels such as the carotid artery is easy and non-invasive. Such experiments could have been tried 60 years ago, when the possibility of a relationship between atherosclerosis and scurvy was first noticed. The delay is unfortunate, because if the hypothesis is correct, then people continue to die unnecessarily.

Vitamin C and lysine

"I think I know what the answer is. I think we can get almost complete control of cardiovascular disease, heart attacks and strokes by the proper use of this therapy... even cure it." Linus Pauling

Linus Pauling claimed to have described a simple cure for one of the major killer diseases of the western world, but was greeted with disbelief.[338,339] Pauling was convinced that heart disease was caused by shortage of vitamin C in the diet, so he devised what became known as the Pauling therapy. Although the explanation for this therapy is based largely on the action of ascorbate, the Pauling therapy also included the amino acids lysine and proline, which are normal components of the diet. Furthermore, a form of cholesterol was suggested as an additional cause of atherosclerosis. Inclusion of these secondary factors complicates the mechanism and makes it more difficult to evaluate, while lessening its scientific impact.

Since the death of Linus Pauling, a former collaborator, Mathias Rath, has made claims concerning what he terms a new revolution in medicine. *Cellular medicine* is a collection of treatments that Rath claims will cure heart disease, cancer and other afflictions. His website and recent books suggest that Rath sees himself as a misunderstood pioneer, who has fundamentally changed medicine. These ideas of personal greatness do not encourage evaluation of his ideas by physicians and scientists, who expect a degree of humility in making scientific claims. Most professional scientists know how difficult it is to find out new things about how the world works. It is easy to make a mistake, or to be misled. A person making grand statements needs strong supporting evidence.

The medical profession found it hard to accept Pauling's support for the idea that vitamin C could be used to prevent heart disease. Many considered Linus Pauling to be little more than a "quack". They had been prejudiced against his ideas on vitamin C, ever since publication of his book on the common cold. His subsequent claim that vitamin C could help cure heart disease seemed incredible. On examination, we find that Pauling was unable to test his claims in the rigorous way that the hypothesis demands. This was not through choice; his research proposals were repeatedly denied funding.

Pauling and Rath's explanation for the Pauling therapy is complicated and is not fully supported by the scientific evidence. While Pauling had a history of successful scientific prediction, this should not prevent critical discussion of his ideas. We have simplified a complex area here, because the claims for this therapy are strong and the description normally given is misleading.

The Pauling therapy

Briefly, the Pauling and Rath model can be stated as follows: when there is not enough vitamin C, a form of cholesterol called lipoprotein(a), or Lp(a), is deposited in the arterial wall. The deposition of Lp(a) cholesterol thickens and strengthens the arterial wall and repairs damage caused by shortage of vitamin C. In the longer term, the Lp(a) cholesterol ceases to be helpful and becomes harmful. As the damage continues, chronic inflammation sets in, the Lp(a) cholesterol becomes oxidised and plaques form. The amino acid lysine acts against this process, by preventing Lp(a) cholesterol sticking to the artery wall.

The Pauling therapy for curing atherosclerosis and coronary heart disease	
Nutritional substance	Grams per day in divided doses
Vitamin C	3 – 6
Lysine	3 – 6
Proline	0.5 – 2

Pauling and Rath explained their claim, that vitamin C and lysine can not only prevent but also cure atherosclerosis, in terms of stopping Lp(a) cholesterol bonding to blood vessels.[340] According to their patents,[341,342] a binding inhibitor such as lysine, used alone or in conjunction with vitamin C, dissolves plaques. Although the central idea is that heart disease is a form of long-term scurvy, the introduction of Lp(a) cholesterol into the model has the unfortunate consequence of eclipsing this. The explanation given for the Pauling therapy depends on another form of "bad" cholesterol.[343]

There is reasonable agreement that Lp(a) cholesterol is involved in heart disease. Rath examined 107 patients, who were undergoing bypass surgery, and found accumulations of Lp(a) cholesterol in their aortic walls.[344] Bypass patients (40% of 306) were more likely to have high levels of Lp(a) cholesterol in their blood than normal controls (16% of 72). The total cholesterol and very low-density cholesterol levels in the bypass patients were also raised. Rath also reported that high levels of Lp(a) cholesterol had accumulated in arterial plaques in autopsy material from 74 subjects.[345] Nachman studied 17 male monkeys, and showed that diet-induced arteriosclerosis involves accumulation of Lp(a) cholesterol in atherosclerotic arteries, but not in healthy ones.[346] Based on such data, Rath argued that Lp(a) is the predominant risk factor contributing to the progression of atherosclerotic lesions in man.[347] However, the experiments reported do not uphold his conclusions; at most, Lp(a) is secondary to vitamin C deficiency.

Around the time the Pauling therapy was being formulated, research interest in Lp(a) cholesterol and atherosclerosis was strong.[348,349,350,351] Plaques contain large amounts of cholesterol, including Lp(a), which is deposited in the arterial wall, particularly in diseased areas.[352,353] However, this research does not indicate a special role for Lp(a). The published work shows only that Lp(a) is concentrated in plaques, where cholesterol is known to occur. It has not been demonstrated that it specifically *causes* plaques to form. The proposal for lipoprotein(a) is a minor variation on the "bad" cholesterol hypothesis, and is intellectually unexciting.

Pauling proposed that Lp(a) cholesterol acts as an antioxidant replacement for vitamin C. Lp(a) cholesterol contains a protein which has more than 100 disulfide groups on each molecule, and these could act as antioxidants.[339] Since the protein is located in the outer shell of Lp(a), it could quench free radicals before they reach the cholesterol contained inside. However, the evidence presented for Lp(a) cholesterol acting as an antioxidant is both indirect and speculative. More importantly, it would cease to have an antioxidant action in an oxidising environment, such inside a plaque.

Lp(a) cholesterol is not a causative agent for the development of heart disease. It may be a risk factor,[354,355,356,357,358,359] but the results are conflicting.[360,361] Simons examined Lp(a) cholesterol concentration in 1202 males and 1512 females, aged 60 and above.[362] He reported that 24% of the males and 17% of the females had coronary heart disease at the start of the study. Lp(a) cholesterol concentration was slightly higher in the subjects with heart disease but the difference was not statistically significant. Another scientist, Gazzaruso, measured Lp(a) cholesterol levels in 249 patients with essential hypertension, in 142 non-hypertensive patients with heart disease and in 264 healthy controls.[363] Gazzaruso indicated that high Lp(a) cholesterol levels were a strong predictor of coronary heart disease in people with high blood pressure. Retrospective case studies by Marcovina indicated that raised plasma Lp(a) cholesterol is associated with coronary heart disease.[364]

Several large population studies have failed to confirm that Lp(a) cholesterol is an independent risk factor.[365] However, raised blood levels of this form of cholesterol may increase the threat associated with traditional risk factors. One study claimed that protein fragments from Lp(a), found in urine, were predictive of coronary heart disease.[366] While there is evidence for uptake of LDL and Lp(a) cholesterol by plaque tissue,[367] high levels of cholesterol in the blood do not necessarily cause arterial plaques to form.

Lp(a) cholesterol - a replacement for vitamin C?

The argument for how Lp(a) is involved in heart disease is based on the idea that arterial plaques are caused by a shortage of vitamin C. Pauling and Rath proposed that Lp(a) cholesterol acts as a replacement for ascorbate, in animals that have lost the ability to synthesise the vitamin.[368] They claim that Lp(a) is more common in animals that need ascorbate in the diet, but do not give direct evidence. Instead, they argue that during evolution, low levels of vitamin C in pre-humans would have caused scurvy and fragile blood vessels. Human evolution might have favoured individuals with an alternative method of strengthening their arteries and Lp(a) could have been used in this way. This evolutionary argument is weak and scientifically untestable.

Vitamin C and Lp(a) cholesterol

The main weakness of the Pauling and Rath theory is the explicit emphasis on Lp(a) cholesterol as a fundamental cause of heart disease. The idea that Lp(a) cholesterol acts as a surrogate for vitamin C is consistent with known facts, but not fully convincing in itself. This lipoprotein may be related to heart disease; then again, it may not.

Pauling and Rath postulated that ascorbate can reduce or prevent the development of atherosclerosis by lowering plasma Lp(a), decreasing lipoprotein infiltration into the arterial wall, and preventing lipid peroxidation.[368] They added that the atherogenicity of Lp(a) seems to be closely related to the ascorbate concentrations in plasma and tissue, and suggested that ascorbate deficiency increases plasma Lp(a). Following their evolutionary argument, we can take this to imply that if sufficient vitamin C is present, the Lp(a) would not be required to act as a surrogate for it. An alternative explanation is that the "atherogenicity" is simply related to plasma levels of ascorbate, and only indirectly, if at all, to levels of Lp(a).

In support of the suggestion that vitamin C lowers plasma Lp(a), they published a study of 11 outpatients with coronary heart disease, who were supplemented with nine grams of vitamin C per day. Lp(a) cholesterol in the blood was reduced by 27%; however, the controls in this study were limited.[385] Evidence that vitamin C lowers Lp(a) is mixed. Bostom gave patients with coronary heart disease 4.5 grams per day of vitamin C, for 12 weeks.[369] He found no effect on blood levels of Lp(a) cholesterol. This double-blind clinical trial used a short period of supplementation. The results were not consistent with vitamin C lowering Lp(a) in the blood. In another study, Jenner's group administered a gram per day of vitamin C to 101 healthy men and women, ranging in age from 20 to 69 years, for eight months. They concluded that it did not lower blood levels of Lp(a) cholesterol in healthy subjects.[370] The failure to demonstrate, consistently, that vitamin C supplementation lowers Lp(a) levels tends to refute the suggestion that ascorbate deficiency causes atherosclerosis by increasing plasma Lp(a).

The lysine connection

Pauling and Rath have suggested that lysine could reduce the binding of Lp(a) cholesterol by preventing it attaching to the arterial wall. In order to strengthen the artery and ultimately form a plaque, the cholesterol has to stay in the artery wall. Pauling and Rath assume that Lp(a) cholesterol binds to damaged arteries. There is supporting evidence for this claim,[371,372,373,374,375,376,377,378,379] and the adhesion may be increased with inflammation.[380] A protein in Lp(a) is chemically similar to one that is incorporated into blood clots. There is some supporting evidence for specific binding of Lp(a) in plaques,[381,382,383,340,384,385,386,387] but once again it is incomplete.[349]

The use of lysine to dissolve arterial plaques is an interesting and important idea. However, this concept does not depend on a specific form of cholesterol, so the action of lysine should be considered separately from Lp(a). Rath's questionable introduction of Lp(a) cholesterol into the debate does not invalidate the central hypothesis. Lysine could be a useful treatment, acting to release lipoprotein and other sources of cholesterol from inside arterial plaques.

More importantly, lysine could act on other forms of cholesterol to prevent plaque formation. If lysine prevents attachment, this would lessen the time it takes a lipoprotein particle to pass into and back out of the plaque. The reduction in transit time would reduce the chance of the lipoprotein being oxidised in the inflamed plaque. Even if it did become oxidised, it might leave the area quickly and not contribute to local inflammation. These two mechanisms could combine to prevent plaque formation and help heal existing damage. Both HDL and LDL cholesterol contain lysine residues.[388,389] Modifying the lysine binding sites on LDL cholesterol inhibits specific binding to platelets.[390] Lysine may act generally to prevent cholesterol build up in the artery wall.

The reliance on lipoprotein (a) in the Pauling and Rath model is probably an error. However, there is a more disturbing problem in our review of the Pauling therapy. The reason the therapy is no more than a hypothesis, is that it has never been tested. The therapy has been actively promoted for over a decade, and there are numerous positive reports of self-treatment on the internet and elsewhere. Despite this, we did not find even a single case report, positive or negative, in the medical literature.

Antioxidants and heart disease

"She was wise, subtle, and knew more than one way to skin a cat." Mark Twain

The evidence indicates atherosclerosis is caused by prolonged, low level deficiency of vitamin C and other antioxidants in the diet. Most heart disease and strokes are therefore a form of scurvy. In addition, vitamin E, in a specific form, has been reported as an effective treatment for heart disease. This suggests that suitable antioxidant supplements can both reverse and prevent atherosclerosis.

Tocopherols and tocotrienols

Evans and Bishop discovered vitamin E in 1922, when they found that rats raised on a diet of milk were unable to reproduce. The pups died in the womb. Wheat germ and lettuce were found to remedy the problem and, in the following years, the different vitamin E substances were identified. The name tocopherol derives from Greek words meaning to bring forth offspring.

We have seen that vitamin E is not a single substance, but a complex mixture of chemicals. In an earlier chapter, we described a biased experiment that claimed to show no effect from antioxidant vitamins on heart disease. This clinical trial included the synthetic alpha-tocopherol form of vitamin E but ignored the tocotrienols. It is important to be aware of the components of vitamin E and their differing roles in heart disease. We will now look briefly at the properties of tocotrienols, which are found in rice and palm oils.

Tocotrienols are similar in structure to tocopherols but have a shorter side chain that contains double bonds; this gives them different properties. The tocotrienols exist as optical isomers and come in d- and l- (right- and left-handed) forms. Moreover, there are several forms of tocotrienols, known as alpha, beta, delta and gamma variants. Tocotrienols spread out and disperse within cell membranes, whereas tocopherol molecules tend to clump together. It is claimed that tocotrienols are more effective as antioxidants and more mobile than tocopherols.

Reversal of heart disease

For many years, health specialists have promoted the beneficial effects of vitamin E in heart disease and atherosclerosis. Dr Evan Shute and his brother, Dr Wilfred Shute, became interested in the vitamin in about 1936. They looked into its use as a treatment for heart disease and other problems. In the early days, the poor availability and quality of vitamin E preparations hindered their studies. Nonetheless, they obtained some positive results and backed up their claims with clinical observations on thousands of patients over several decades, together with additional experimental results.[391]

Predictably, the medical establishment attacked both the Shutes and their results. In 1964, Herbert Bailey published a book called "Vitamin E: your key to a healthy heart", in which he described how the medical establishment criticised the Shutes and suppressed their findings. Herbert Bailey wrote his book after having a heart attack that he treated with vitamin E, following the Shutes' advice. Over the decades since the Shutes' original work, there have been many studies of the effect of vitamin E on heart disease. While the results have not all been positive, the evidence supports a protective effect, especially for its use in combination with vitamin C.

The Shutes' work has been partially confirmed but the treatment is now thought to be less effective than originally suggested. The Cambridge Heart Antioxidant Study,[392,393] or CHAOS, was an intervention based clinical trial, in which 2002 heart patients were divided randomly into three groups and given either 400 IU or 800 IU of vitamin E daily, or a placebo, for 200 days. Both vitamin E groups showed a 77% decrease in levels of non-fatal heart attack (14 out of 1035 vitamin E subjects, against 41 out of 967 subjects that were not supplemented) but there was no significant difference in the number of deaths. Losonczy examined vitamin E and vitamin C supplementation in 11,178 elderly people aged 67-105 years, from 1984-1993.[394] Simultaneous use of vitamins E and C was associated with a lower risk of death from all causes, and fewer deaths from coronary heart disease.

The well-known Harvard studies indicated that supplementation with more than 100 IU of vitamin E, for more than two years, decreased heart disease by 40%.[395,396] Subjects were 87,245 female nurses, aged 34 to 59 years, in whom 552 cases of major coronary disease were noted over an eight year study period, and 39,910 U.S. male health professionals, aged 40 to 75 years, in whom 667 cases of coronary disease were observed over a four year period. The subjects had no previous history of heart disease. The male study also showed a positive effect for vitamin C, but this was not significant at the doses investigated.

The central importance of vitamin E is illustrated by the work of Fred Gey, who indicated that the most important risk factor for heart disease was a deficiency of vitamin E.[397,398] In each of the 12 countries he studied, vitamin E appeared to be of greater significance than a combination of smoking, blood pressure and blood cholesterol.

Reversal of atherosclerosis

In 1992, Dr. Anthony Verlangieri, of the Atherosclerosis Laboratory at the University of Mississippi, demonstrated the use of antioxidant supplements to prevent and reverse atherosclerosis in monkeys.[399,400] Shortly after, studies with humans indicated that antioxidants could slow the progression of atherosclerosis.[401] DeMaio studied one hundred patients, who had undergone coronary angioplasty to stretch atherosclerotic blood vessels in the heart.[402] The patients were given either a large dose of vitamin E or a placebo daily. Four months later, restenosis or re-narrowing of the arteries, because of build-up of plaque, was found in about half the placebo subjects but only one third of those supplemented with vitamin E.

Using the right kind of vitamin E

Vitamin E can reverse heart disease. Tocotrienols have been shown to reduce the arterial blockage caused by plaque and arteriosclerosis. In a three year, double-blind clinical study, Dr Marvin Bierenbaum showed that tocotrienols can lead to regression of atherosclerosis.[403,404] His research group had observed that tocopherols and tocotrienols greatly reduced activity in the platelets and could prevent clots forming. Platelets are small, disc shaped particles in the blood, made in the bone marrow. They are an essential part of clot formation and accumulate in large numbers at the site of an injured blood vessel.

Since tocotrienols had been shown to reduce the effects of dietary cholesterol in rats, Bierenbaum decided to try them in humans. In 1995, Bierenbaum's research group showed that tocotrienols reduce blood vessel blockage. They measured the effects of both gamma-tocotrienol and alpha-tocopherol on carotid atherosclerosis. This study estimated the degree of carotid artery blockage, over 18 months, in 50 patients with cerebrovascular disease. Ultrasound scans of the artery were done at six months, twelve months, and continued yearly.

Initially, tocotrienols reversed the disease in a proportion of the patients. The ultrasound measurements revealed improvement in seven of the 25 patients given gamma-tocotrienol, and progression in two. None of the control group of 25 patients improved but 10 showed increased disease and blockage. Blood platelet oxidation decreased in the treatment group after 12 months but the placebo group showed an increase, although it was not significant. Blood levels of LDL (bad cholesterol), HDL (good cholesterol) and triglycerides remained unchanged. This implies that plaques can be reversed independently of the level of blood cholesterol. The effect of the tocotrienols increased with time and, ultimately, the tocotrienols provided benefits in almost all patients. In a later report, 92 percent (23) of the patients in the tocotrienol group stabilized or improved. In the control group, 48 percent (12) of the patients deteriorated and no patient improved.

Bierenbaum suggests that oxidized cholesterol exists in both stable and mobile pools within an arterial plaque. Tocotrienols prevent oxidation of the cholesterol in the labile pool, and help move it out of the artery wall before it becomes attached to the plaque. He argued that this mechanism would explain the dramatic effects for patients supplemented with tocotrienols, which gave results suggestive of a cure after only six months of treatment. This explanation for the reversal of plaques with tocotrienols depends on their action as local antioxidants. It is consistent with the antioxidant effect described for vitamin C. Indeed, Bierenbaum's explanation, that tocotrienols prevent cholesterol molecules sticking in plaque, is similar to the action of vitamin C and lysine proposed by Linus Pauling.

Animal studies

Small animal studies have confirmed the potential benefits of tocotrienols on heart disease. Certain mice, with a genetic defect in lipid processing, get atherosclerosis only when fed a high cholesterol diet. A study of such mice found that tocotrienols inhibited plaque formation.[405] Unsupplemented mice had large plaques at the level of the aortic valve. Mice given a lower dose tocotrienol supplement had plaques that were 92% smaller. The plaques of those receiving a higher dose supplement were 98% smaller. In other words, the control mice had large plaques and the treated mice had almost no plaques at all. This difference in plaque size occurred despite the treated mice having similar blood fats to the controls. A second research group, working on a similar strain of mice, confirmed and extended these results.[406]

In a further study, 18 rabbits, in three groups, were fed an unnatural, cholesterol rich diet for 10 weeks. The results suggested that tocotrienols could have beneficial effects on blood lipids and could reduce oxidation in arteries.[407] A second study, carried out on rabbits over a 12-week period, indicated that tocotrienols provided greater protection against plaque formation than the more common tocopherol form of vitamin E.[408]

Mechanism of action

Two independent lines of research, based on two different vitamins, indicate that antioxidants can reverse atherosclerosis. Why do these particular antioxidants work when others, such as the tocopherol forms of vitamin E, do not? A likely explanation is that they are mobile molecules, which can penetrate the plaque in sufficient quantity to have the desired effect. Conversely, while supplements of many other antioxidants accumulate in the blood, they do not appear to enter plaque at high levels and are therefore ineffective.[409]

At first glance, tocotrienols appear to work better than vitamin C. This may relate to the lack of experimentation with high dose ascorbate. Bierenbaum's clinical study with tocotrienols covered a longer period than the six-month studies by Willis and Rath on vitamin C and plaques. Vitamin E is fat-soluble and stays in the body for longer. It remains possible that larger, divided doses of vitamin C, over a longer period, will be more effective than tocotrienols. However, since both vitamin C and tocotrienols are thought to operate by the same antioxidant mechanism, there exists the exciting possibility that they could be synergistic.

Given these findings, why did the world's media not immediately announce the curative potential of tocotrienols? In an interview with the nutritionist Dr Richard Passwater, Bierenbaum was asked why he did not make more of his potentially important results.[400] His reply indicated that this was the age of the mega-study and the meta-analysis of results from several large studies. Bierenbaum's study included only 50 patients and he suggested it was considered less solid evidence than a mega-study. If this really is the case, medical science has lost its way. To reiterate, validation comes from replication. If you are looking for a reliable cure for heart disease, you will find it with a small study, like that described by Bierenbaum, followed by replication. An alternative explanation for the absence of follow up studies is that pharmaceutical companies could not make money out of these results and would lose billions of dollars in drug sales if they were confirmed.

Vitamin C and tocotrienols can reverse coronary heart disease. An important question is whether high dose vitamin C and the tocotrienols, if supplied together, would work in an antioxidant network. If so, a combination of high dose vitamin C and tocotrienols would be appropriate for removing arterial plaques. On the available evidence, this combination could be curative and has no known harmful effects.

Lipoic acid

Lipoic acid is a small, mobile antioxidant, which is soluble in both water and oil. It potentiates many of the effects of vitamin C. Because of its chemical properties, we would expect this substance to enter arterial plaques. It is a powerful antioxidant and may provide protection against heart disease.[410] Lipoic acid is part of the antioxidant network and can reduce other antioxidants, enhancing the function of ascorbate and acting as a substitute for vitamin E.[37,411]

Lipoic acid has demonstrated a preventative effect on the development of plaques, in a study of the Japanese quail.[412] This little bird (Coturnix coturnix japonica) may be a useful laboratory animal for the study of atherosclerosis, as it is omnivorous, easy to maintain, and susceptible to spontaneous and cholesterol-induced atherosclerosis. Some birds have a reduced capability for making endogenous vitamin C,[47] although Japanese quail do not have a specific requirement for it as an essential nutrient. However, supplementation with vitamin C reduces the minimum requirement for the antioxidant riboflavin (vitamin B2) by half, indicating that endogenous production may be inadequate.[413,414] The quail has a short life cycle and develops human like plaques in the aorta and sometimes in the coronary artery. Subcutaneous implantation of slow release lipoic acid inhibited the development of such plaques.

In humans, lipoic acid changes the composition of blood fats in a way that could provide protection from heart disease.[415] It can inhibit the attraction of white blood cells to inflammation more powerfully than aspirin,[416] and hinders one of the first steps in plaque formation, the adhesion of white blood cells to the arterial wall.[417] Lipoic acid stabilises mitochondria, reducing superoxide formation.[44,45,46] It may decrease the oxidant effect of glycated proteins in arterial plaques, although the effect is smaller than that of vitamin E.[418] The anti-inflammatory effect of lipoic acid may result from its antioxidant activity,[419] as well as by inhibition of an important regulator of gene expression.[420] Lipoic acid is also involved in cell signalling and cell death by apoptosis.

Lipoic acid is generally considered a safe food supplement. There is no direct evidence to suggest that lipoic acid will reduce plaque size in humans, as the clinical studies have not been performed. If, however, the aim is to provide a reducing state inside plaque tissues, adding lipoic acid to the diet can be justified on biochemical grounds.

Antioxidant network therapy

From the current evidence, we can suggest a modification to the Pauling therapy, which we will refer to as Antioxidant Network Therapy. The evidence for the involvement of lipoprotein(a) in Pauling's therapy is not as strong as the support for additional antioxidants, such as the tocotrienols. In this book, we are explicitly not promoting any therapy or suggesting supplementation; there are many self-help books of that nature available. However, we do need to provide an indication of the kind of therapy most likely to achieve the stated aim, for future research purposes. In this case, the Pauling therapy is deficient and here we correct those aspects of the published suggestions.

Prevention

Prevention of heart disease typically requires about three grams of vitamin C, spread throughout the day, to ensure that deficiency does not occur. In general, this should be taken with a multivitamin and a broad spread of other dietary antioxidants, such as vitamin E and coenzyme Q10. A person at particular risk might consider doubling this level of vitamin C and adding 250mg of mixed natural tocotrienols combined with extra tocopherols, and r(+)lipoic acid. (The r(+) form of lipoic acid is more physiologically available and to avoid future experimental bias this needs to be stated explicitly.) Including lysine and proline as suggested by Pauling seems to be justified, as they have low toxicity and may be beneficial. This regime for prevention is similar to the therapy proposed by Pauling, with the addition of further antioxidant support in the form of tocotrienols and lipoic acid.

Treatment

To get rid of existing atherosclerotic plaques, drastic action is needed. Pauling's suggested doses are probably too small for clinical benefit. Cathcart's bowel tolerance method might be more suitable. The resulting vitamin C dose will be variable but more likely to be appropriate to personal biochemistry. In some cases, intravenous ascorbate might be needed.

In cases of existing disease, at least six months of treatment with either antioxidant network therapy or the Pauling therapy would be needed, before benefits in the form of plaque reduction are likely to become apparent. We would expect to see plaque reduction within one year of starting the antioxidant network treatment, and a maximal effect within two years.

	Pauling therapy	Antioxidant network therapy
Vitamin C	3 - 6 grams	At or near bowel tolerance: 6+ grams
Lysine	3 - 6 grams	3 - 6 grams
Proline	0.5 - 2 grams	0.5 - 2 grams
Tocotrienols	-	300+ mg
Tocopherols	-	800+ IU
r-Lipoic acid	-	300 - 600 mg

The contraindications are those we have described previously for high dose supplementation of vitamin C. Additional safety considerations apply with the other supplements such as high dose lipoic acid. There may be further factors for an individual and, as with any treatment, this regime should only be considered with medical advice.

Why has antioxidant therapy not been properly investigated?

The evidence we have considered in this book is not conclusive. It does not "prove" that vitamin C or, for that matter, vitamin E as tocotrienols, will prevent or cure atherosclerosis. Such claims and objections are not scientific.

Linus Pauling suggested his treatment could even cure the condition. The medical profession tends to avoid the word "cure", as it can be too strong a claim. As we have explained, insulin treatment does not cure diabetes but converts it to a long-term degenerative condition. Critics point out that one reason for medicine avoiding the word cure is that curing a chronic condition is not in the financial interests of pharmaceutical and related industries. A treatment for a chronic condition is more profitable than a cure, as it continues indefinitely.

Antioxidant therapy may indeed be curative, as Pauling and others have suggested. We do not know if this is in fact the case and neither do the medical profession. More importantly, they appear to have little interest in finding out. It is almost unbelievable that medical science has ignored theories of the cause and possible cures for the biggest killers in the western world. The theories have been proposed by reputable scientists over a period of half a century.

When we looked at James Lind's cure for scurvy, the question was, would you eat the citrus fruit if you were on the ship and knew the results of the experiment? The current question is essentially the same. If heart disease and stroke could be caused by insufficient vitamin C in the diet, is there any reason not to take supplements?

Infectious diseases

"We have to date not had a single death amongst our patients with full blown AIDS who have continued on our vitamin C and nutrition programme." Dr Ian Brighthope

The claim that vitamin C can prevent and relieve the symptoms of the common cold seems a relatively insignificant suggestion. However, around half the total number of visits to family doctors are for colds, flu and bacterial infections.[421] This implies a significant financial impact on health budgets. Results to date suggest that vitamin C can play a role, both in preventing colds and in limiting the symptoms. Research also supports a wider role for vitamin C in the control and treatment of infectious disease.

The evidence for the use of vitamin C in infections is substantial. Recently, Dr Thomas Levy has reviewed the literature and described his own experiences in his book "Vitamin C, infectious diseases and toxins".[422] He provides a comprehensive listing of the antiviral, antibacterial and antitoxic effects of vitamin C. Levy reports over 1200 studies, supporting the case that vitamin C can cure infectious diseases. Reports from other physicians, including Fred Klenner, Robert Cathcart and Archie Kalokerinos, describe findings from thousands of patients, treated over the last few decades.

As readers will now suspect, the studies on patients receiving low doses of vitamin C give conflicting results. These studies have no relevance to the large doses that, according to the dynamic flow model, are required for treating disease. Firstly, they have ignored the fact that the half-life of vitamin C in the blood is only about 30 minutes. Secondly, they have used too small a dose. Double-blind clinical trials, using high dose vitamin C as a treatment for infection, do not appear to exist. While there are no positive double-blind clinical trials, there are also no refutations. The failure to carry out the necessary experiments reflects prejudice at the core of medicine.

Normally, the absence of double-blind clinical trials would be considered a severe limitation on the evidence. However, an overriding factor makes the risk of experimenter bias and placebo effects less important. When infections are treated with massive intravenous doses of sodium ascorbate, the reported effect is large: on a par with the effects of antibiotics before the development of bacterial resistance. Experienced physicians report that, with sufficiently high doses, treatment of most infectious disease results in remission. These substantial effects have been repeated by a number of independent investigators. Such striking observations cannot be explained away by the suggestion that they were not made in a double-blind clinical trial, or by attributing them to an error in observation. Criticism of this form is inherently unscientific; if there is any doubt, the experiments should be replicated.

Over a period of at least half a century, numerous independent physicians have made consistent claims for the effects of large doses of vitamin C in fighting infection. This means that if the doctors performing these studies are not all repeating the same mistake, then they are making regular observations of fact. Since there is no contradictory evidence in the literature, we have to take these reports as indicating a genuine effect. As we shall see, Fred Klenner describes treatment of 60 cases of polio that recovered uneventfully over a few days. Ian Brighthope and Robert Cathcart have each reported cases of full-blown AIDS going into remission.

The consistency of the response increases with the dose given. This is what we would predict, based on the dynamic flow model described earlier. Thomas Levy suggests that just as the three most important factors in buying a house are said to be "location, location and location", the three most important factors when considering the action of vitamin C in disease are "dose, dose and dose".

Early Studies

Early studies of the clinical effectiveness of vitamin C in the treatment of infection do not meet current standards of scientific rigour. Nonetheless, they are interesting, as they provide the background to more recent work. The relationship between scurvy and other diseases was noticed long ago. Richard Morton's classic work "Phthisiologia", dated 1689, includes the statement that "scurvy is wont to occasion a consumption of the lungs", suggesting a relationship between vitamin C deficiency and tuberculosis.[422] However, studies of interactions between vitamin C and infectious diseases could not begin properly until after 1932, when the vitamin was isolated and identified.

Since 1935, it has been known that vitamin C can inactivate some viruses.[423] In 1936, sufficient vitamin C became available for an initial study of whooping cough, by Otani in Japan.[424] The dose was low, 100mg by injection, presumably because the newly purified vitamin was in short supply. This uncontrolled study of 81 patients indicated clinical benefit.

The vitamin appeared to inhibit the growth of the pertussis bacteria that causes whooping cough, although it was ineffective with a range of other disease causing bacteria. Over the next two years, more case studies of the successful treatment of whooping cough with vitamin C were reported.[425,426,427]

In the 1940's, the first large dose studies suggested that vitamin C was an effective treatment for colds.[428] A year later, Dr Fred Klenner reported that intravenous doses of one gram of vitamin C were effective against viral pneumonia and polio.[429,430] Even at this early stage, Klenner noted that most clinical studies on vitamin C had used too small a dose. In 1950, a case report on seven patients with rheumatic fever indicated that four grams of vitamin C was beneficial, and added that high doses were harmless.[431]

In 1951, Klenner indicated that massive doses of vitamin C were effective against viral diseases.[432] He claimed that the action of the vitamin against viruses is proportional to its concentration and to the duration of the treatment. In addition to its antiviral action, he reported that vitamin C activated the body's immunological defence mechanisms. He illustrated his assertions with several case reports. In the same year, McCormick hypothesised that increased vitamin C intake, from citrus fruits and tomatoes, was largely responsible for the falling death rate from infectious diseases.[433] He later stated that vitamin C had a broad-spectrum antibiotic action against a wide variety of viral and bacterial diseases.[434]

By 1957, Klenner was suggesting that vitamin C should be given for all infectious diseases, while the doctor considers the diagnosis.[435] He practiced this approach for many years with apparent success.[436] Reviews of the early studies of vitamin C conclude that they have demonstrated direct antiviral activity against a wide range of pathogens.[437,423] In 1969, Klenner reported that, in his clinical experience, ascorbate in sufficiently high doses could clear any viral disease within 24 hours, an apparently outlandish claim by modern medical standards.[438]

In the period following the initial isolation of vitamin C, a number of studies showed it to have both direct and indirect antiviral activity. Vitamin C has been shown to have antiviral properties against the cold virus,[439] herpes simplex, cytomegalovirus (a form of herpes) and an influenza-like virus, especially in the presence of copper.[440,441] A modified long-acting form of vitamin C also has antiviral activity against several types of human cytomegalovirus.[442] Interest in the antiviral effects of vitamin C lasted for about twenty years following its isolation, but seems to have diminished in the 1950s, after the introduction of antibiotics and immunisation. However, antibiotics are not effective against viruses, so it surprising that attention should have waned at this time. Perhaps vitamin C was overshadowed by the hubris following the successful introduction of antibiotics and immunisation campaigns.

The proposed general action of vitamin C against infections cannot be explained as an antiviral or antibacterial action. Vitamin C will inhibit some pathogens but not others. Indeed, it is possible to grow bacteria using ascorbate as a source of energy. If vitamin C has the effect of increasing the body's general resistance to infection, it acts by strengthening the host response. It could achieve this effect either by a direct action, dependent on its antioxidant properties, or by influencing the host response through redox signalling. If this is the case, the effect must be substantial.

The common cold

The common cold has been a focus of controversy about the effectiveness of vitamin C as a treatment. There is evidence that large doses can prevent colds, but response to oral vitamin C as a treatment appears variable. The continuing argument about vitamin C and the common cold is unfortunate, as colds are minor infections, of peripheral medical interest. To be clear from the outset, people taking megadoses of vitamin C will still get the common cold and other infections from time to time. The frequency of such illness will be lower and the symptoms may be less severe. Gram level oral doses of vitamin C are not an effective treatment for the common cold.

A one-gram vitamin C tablet is unlikely to have anything but a small effect on the progress of a cold; both the sceptics and the proponents of ascorbate agree on this. Occasionally, sceptics incorrectly suggest that a gram level dose has been claimed to cure a cold, adding that if it has no effect, the proponents must be wrong. Larger doses, say 5 or 10 grams, may relieve a cold but the effects will be variable and many people will not benefit. Cathcart's threshold theory of the action of vitamin C may originate from experimentation with symptoms of the cold.

Titration to bowel tolerance level

According to Cathcart, to relieve the symptoms of a cold, you should titrate to bowel tolerance. Titration is a term used in chemistry, when you gradually add one chemical to another, quickly at first, then drop-by-drop, until there is just enough to initiate a reaction. Titration to bowel tolerance might entail taking a gram or two of vitamin C dissolved in water every half hour, in a trial and error process. An initial loading dose of perhaps five grams might be accepted and reduce the time needed to reach gut tolerance level. The dosing continues until the first indications of gut irritation are felt, for example, a small amount of gas or rumbling, and then smaller doses are given. Too high a dose causes diarrhoea, which is a signal to reduce the dose level slightly.

Readers can experiment with the titration technique for themselves, when they notice a cold coming on. Titration to bowel tolerance may take a little practice, but it is possible for an individual to investigate Cathcart's claims and conduct their own case study on the common cold. Although interesting, the self-conducted experiment has limited value in assessing the utility of vitamin C in infectious disease, because of the placebo effect and biological variability. Some proponents of vitamin C have made great claims for this approach, but other people suggest it does not work.

We have explained previously why variation in the results is to be expected. People with less severe colds may reach high enough levels to quench the inflammation, while others with infections that are more virulent do not. Over the last 50 years, the literature has consistently reported that intravenous doses are far more effective in the treatment of acute infections. Intravenous sodium ascorbate produces much higher blood levels than oral doses. However, intravenous infusions are not given for minor infections like the common cold.

Dose levels and the threshold effect

The misconception that vitamin C is ineffective as a treatment for colds arose from a review by Thomas Chalmers in 1975.[443] This study was influenced by Linus Pauling's assertions that vitamin C could prevent the common cold and other infections. Chalmer's review was negative and did not agree with the results of others, who indicated a positive effect. Dr. Harrie Hemila reanalysed Chalmer's study and noticed that a study of only 25-50mg had been included and had biased the results. No one was suggesting that so low a dose could be beneficial. Chalmer's error reinforced the idea that vitamin C was ineffective.

A recent review of 30 published studies indicated that colds were not prevented by doses in the region of one gram daily of vitamin C, but found that this dose did reduce the duration of illness.[444] A further study reports a reduction of 80% in the number of cold sufferers supplemented with vitamin C who went on to develop complications such as pneumonia.[445,446] Hemila indicates that, despite widespread scepticism, vitamin C is effective against colds,[447,448,449,450] but suggests that Linus Pauling was over-optimistic about the size of the benefit.[451,452]

In 1999, Gorton and Jarvis studied the effects of vitamin C on cold and flu symptoms.[453] This was a prospective, controlled study of students in a technical training facility. The control group of 463 students were aged from 18 to 32 years. The experimental group of 252 students were aged from 18 to 30 years. The researchers tracked the number of reported symptoms for the treatment group and compared them with those from the control group in the previous year. The control group, on reporting symptoms, were treated with pain relievers and decongestants. The test population had hourly doses of one gram of vitamin C for the first six hours after developing symptoms, followed by one gram three times daily. Symptomless subjects in the test group were also given a gram of vitamin C three times each day. Reports of flu and cold symptoms in the vitamin C group were 85% lower than in the control group. Vitamin C in large repeated doses relieved and prevented the symptoms of colds and flu. However, the controls were less than adequate as, for example, the two populations may have been subject to different viruses.

Prevention and reduction of severity

In 2002, a placebo controlled, double-blind clinical trial demonstrated the effectiveness of vitamin C in the prevention of the common cold.[454] The subjects taking ascorbate had fewer, less severe colds and recovered more quickly. One hundred and sixty-eight volunteers were divided randomly into two equal groups, which received either a placebo or a vitamin C supplement. Both groups were given two tablets daily over 60-days in the English winter. A five-point scale assessed current health and a diary recorded any common cold symptoms. The vitamin C group had 37 colds and this was significantly fewer than the control group, who suffered 50. The ascorbate group also had fewer days of active cold, 85 compared with 178 in the controls. The average duration of symptoms in the treated group (1.8 days) was less than in the controls (3.1 days).

Negative results

Not all recent studies are positive, however. Robert Douglas, from the National Centre for Epidemiology and Population Health in Canberra, Australia, accepted that large prophylactic doses of vitamin C could reduce the severity and duration of colds, but thought it did not affect the incidence. He believed that previous studies on large doses of vitamin C to treat colds were inconclusive and decided to perform his own experiment.[455]

Douglas's double-blind clinical trial, using doses of one to three grams of vitamin C, found no reduction in the severity and duration of colds. Of 400 volunteers in the study, 149 returned records covering 184 cold episodes. No significant differences were found in any measure of cold duration or severity with the dose of vitamin C. The Douglas study was reportedly designed in a statistically sensible way, to find an effect of the vitamin. Robert Douglas said that his study was *definitive* and that he stopped taking vitamin C supplements as a result.[456] It turns out that Professor Douglas's study was not as definitive as he might have thought.

Douglas designed his study to test the effectiveness of vitamin C acting in the same way as a drug. An aspirin taken during a cold will give minor relief from some symptoms. In Douglas's study, subjects were told not to take the vitamin until at least four hours had elapsed since the onset of symptoms. The doses were not repeated and the subjects were told they must have at least two cold symptoms before taking the vitamin. The symptoms were sore or scratchy throat, nasal congestion or discharge, headache or stinging eyes, muscle aches, fever, or "four hours of certainty that a cold is coming on". After waiting until the cold was established, subjects took the vitamin C for a total of three days.

The main deficiency of Douglas's cold treatment study was that it missed the point; large nutritional doses of vitamin C are used to quench a cold *before the symptoms have taken hold*. In his study, the time from the start of symptoms to the beginning of treatment was estimated to be more than 10 hours. If the patient waits several hours until the disease has taken hold before starting the treatment, the appropriate remedy is titration to bowel tolerance. In other words, the sick person needs a pharmacological dose, not a nutritional supplement. Douglas's study results are predicted by the dynamic flow model. The four studies cited by Douglas as giving inconclusive results suffer the same basic shortcoming as his own study.[457,458,459,460] They do not conform to the mechanism by which vitamin C is thought to operate.

Importance of dose

The position with vitamin C and the common cold is becoming clear. Frequent, large nutritional doses can reduce the incidence of colds, their severity and duration. Once a cold has taken hold, treatment is much more demanding. To quench the symptoms of an established cold, the recommended vitamin C treatment is titration to bowel tolerance, which may involve up to 100 grams per day. This is quite challenging and, for a cold, the effort may often not be worth the relief provided. It may be that intravenous infusion of sodium ascorbate would quench cold symptoms in most subjects but such an invasive procedure cannot be recommended.

Influenza

While the common cold is a minor infection, influenza is not. As with a cold, the influenza virus probably starts by infecting the nasal passages and upper airways. The flu victim usually recovers in a week or so, but vulnerable individuals, including the elderly, can suffer complications. Influenza can become a more deep-seated infection, as the virus infects the bronchi and lungs causing bronchitis or pneumonia. The virus replicates and destroys epithelial cells in the bronchi, bronchioles and alveoli, and fluid seeps into the lungs. Then bacteria such as staphylococci and streptococcus pneumonidae invade the tissues, taking advantage of their compromised state. In the UK alone, it is common for influenza to cause 1900 deaths in a single year. It is a misconception that these deaths are all old people; half the victims are young, previously healthy individuals.[461]

While influenza epidemics are common and have a considerable death toll, the virus also has a pandemic face.[462] Every so often, the influenza virus explodes in a worldwide pandemic, killing huge numbers of people. The last major one was the forgotten plague of 1918. This came just after the First World War and killed at least 50 million people. People remember the dead of the Great War, but the greater number of deaths from influenza is largely forgotten.

Minor pandemics occurred in 1957 and 1968, and it is thought likely that another one could arise at any time. The advent of air travel and a more mobile population makes the rapid spread of the disease likely. Although there is surveillance for outbreaks of influenza, especially in the Far East, an outbreak could occur almost anywhere. The 1918 pandemic probably started in Europe, in either Spain or France. Currently, protection against influenza is by immunisation and by some specific antiviral drugs: amantadine, rimantadine and the neuraminidase inhibitors. These treatments can be effective in preventing the spread of infection and are used therapeutically to relieve symptoms.

The literature on treatment of flu using vitamin C is limited. Levy reported a 1963 study by Magne, on treatment of 130 cases given variable doses of up to 45 grams.[463] Of these subjects, 114 recovered well in three days, while 16 did not respond. Levy explained the lack of reaction in a small number of subjects as being due to a biological response to the variable doses administered.

A recent Bulgarian study, in mice, sought to find the effect of injections of vitamins C and E on free radical diseases, particularly infection with influenza.[464] The study indicated that vitamin E reduced lipid peroxidation during the infection. Vitamin C showed a similar, smaller effect but potentiated the action of vitamin E. This study suggested that vitamin C acted by chemically reducing the oxidized vitamin E, thus increasing its effects. We can use the dynamic flow model to help understand the results of this study. Firstly, we point out that mice manufacture ascorbate internally, so injecting it may have the effect of suppressing production. More importantly, a single injection of vitamin C may be excreted before its effects can be seen. With water-soluble vitamin C, in a large animal like man, half an injected dose will be excreted by the kidney in about 30 minutes. In an animal the size of a mouse, a single injection would probably be excreted in a few minutes, and would not be expected to provide protection against infection.

Poliomyelitis

Polio is a disease of sudden onset, with symptoms of a systemic infection, similar to flu. The symptoms either go away quickly or develop further, affecting the reflexes and producing muscle paralysis. While it affects people of all ages, it is generally a childhood illness. Polio is a disease of the entire nervous system but is notable mainly for the paralysis produced by damage to the motor neurons of the spinal cord.

In 1937, Jungeblut reported positive results in treating polio infected monkeys with vitamin C. He used 400mg of intravenous ascorbate, after having infected monkeys with poliovirus using a relatively mild droplet method.[465,466] He also showed that vitamin C could inactivate the poliovirus directly in the test tube.

Jungeblut's colleague, Albert B. Sabin, who later became famous for developing the live oral polio vaccine, disagreed with him. In 1939, Sabin reported that increased vitamin C did not decrease the susceptibility of monkeys to polio. However, he used tiny amounts of ascorbate, employing doses of up to 150mg in his experiments. Furthermore, unlike Jungeblut, he injected the poliovirus directly into the monkeys. Klenner quotes Sabin as reporting, "one monkey was given 400mg of vitamin C for one day at the suggestion of Jungeblut, who felt that large doses were necessary to effect a change in the course of the disease".[467,468] Sabin apparently took little note of Jungeblut's findings, and produced a negative report on vitamin C. The polio vaccine was ready for testing by 1954.

Sabin could have designed his experiments to minimise any positive effect. Certainly, he would not be alone in giving too low a dose of vitamin C and selecting experimental procedures to lessen any observable benefit. This often-repeated manoeuvre could be partly explained by the prevailing idea of a micronutrient. However, it is characteristic of studies that have hidden motives. Sabin's primary aim was to find a suitable vaccine. Because of his work, we now have an improved vaccine, helping us avoid the ravages of polio in the population. While Sabin's work deflected interest away from ascorbate, at least it gave a clear field for acceptance of his vaccination.[467]

In 1949, Dr Fred Klenner reported that vitamin C was an effective treatment for polio and other infectious diseases.[469] Klenner acknowledged that Jungeblut had already demonstrated antitoxic and antiviral properties of vitamin C. He said of other vitamin C studies that "The years of labor in animal experimentation, the cost in human effort and in "grants", and the volumes written, make it difficult to understand how so many investigators could have failed in comprehending the one thing that would have given positive results a decade ago. This one thing was the *size* of the dose of vitamin C employed and the *frequency* of its administration."

Klenner had previously observed dramatic benefits from the use of vitamin C in atypical pneumonia and this led him to try it against other viral infections. He reported results on measles, mumps, chickenpox, herpes zoster, herpes simplex, influenza and atypical pneumonia, but concentrated on polio, as he thought the mechanism of action was similar in each case. He found frequent administration of massive doses of ascorbate to be so encouraging that he decided to review the literature. One study reported lower levels of vitamin C in the urine of supplemented polio sufferers. This suggested to Klenner that the vitamin was being used up by the infection, especially when greater loss was found in more severe disease.

Klenner treated 60 cases of polio with large doses of vitamin C; every patient recovered uneventfully, within three to five days. The patients had the classic clinical signs of polio. The diagnosis was confirmed by lumbar puncture in 15 subjects and eight subjects had contact with a proven case. Vitamin C was given every two to four hours by injection. All patients were clinically well after 72 hours. Three patients relapsed, whereupon vitamin C was restarted and continued for at least another two days, with injections of one to two grams every eight to 12 hours. Where spinal taps were performed, it was the rule to find the spinal fluid returned to its normal, clear state after the second day of treatment. For patients treated at home, the schedule differed and included oral supplementation, as well as injection.

Fred Klenner's work on polio is sometimes described as having "proved" that vitamin C would cure the disease. This is not the case. Klenner reported clinical observations on a number of patients probably suffering from polio. While this, and his reports on other diseases, is interesting, it was not followed up by double-blind clinical trials or experimental studies. In the case of polio, this can be explained partly by Sabin's success with polio vaccination as a preventative measure. Although it is possible that Klenner's observations were flawed and that his results are misleading, no-one has shown this to be true. If we wish to deny Klenner's results, the only credible explanation is that his patients were not suffering from polio at all but had a different disease, such as a mild flu, which somehow made the spinal fluid cloudy and resolved in three days.

In certain cases, it is possible to mistake other diseases for polio. In India in 1993, Dr Ramar reported a four-year-old boy who had been unable to walk for a period of 10 days.[470] The problem had arisen suddenly and acute paralytic poliomyelitis was diagnosed. The boy had a fever, respiratory infection, muscle tenderness, and weakness in the legs. He had not been immunized against polio. One week later, the boy had swelling, thickening, and tenderness of left femur and an X-ray showed features of scurvy. The child responded well to vitamin C and was walking again within two weeks. This report illustrates how a lack of vitamin C, to the levels of scurvy, can result in a pseudo-paralysis that can be mistaken for polio. So, did Klenner's patients really have polio?

It is just about feasible to suggest that Klenner's diagnosis was wrong, in all 60 patients, during an epidemic of polio. If Klenner's observations on polio and other diseases were unique, they would be so out of the ordinary as to be considered unreliable. However, Dr Edward Greer later replicated Klenner's work and reported good results. Greer reported five case studies, where vitamin C was apparently effective in treating polio.[471] Greer was typically giving 10 grams of vitamin C every three hours or with each meal. It is unreasonable to discount studies such as Klenner's based on prejudice; it is necessary to perform the experiment. Polio is not completely eradicated, despite widespread use of vaccination, and outbreaks still occur. Doctors still consider poliomyelitis an untreatable disease, and ignore the potential of ascorbate as a treatment.

AIDS

Acquired Immune Deficiency Syndrome (AIDS) is believed to be the outcome of Human Immunodeficiency Virus (HIV) infection. People with AIDS often suffer lung, brain, eye and other organ disease, along with weight loss, diarrhoea, candidiasis, dementia, toxoplasmosis and a type of cancer called Kaposi's sarcoma. AIDS is a controversial subject; almost everything we "know" about it is disputed. AIDS is also big science and researchers have received billions of dollars to solve the problem, as a matter of urgency. The unfortunate result is that a lot of AIDS research has been of a low standard. Since vitamin C research is itself contentious, we will avoid the AIDS controversies and stick to our core aim.

In 1985, Dr Cathcart had an AIDS patient with an estimated T helper cell (a type of white blood cell) count of five, which is low. This meant that his immune system was severely compromised. The patient's doctors had given him about two months to live. Cathcart administered 60 grams of intravenous ascorbate and advised the patient to take up to 200 grams of ascorbate orally each day. In three months, his T helper cell count was over 500. At the time of writing, he is still alive, HIV positive but with no sign of disease.[58]

In 1987, Dr Ian Brighthope published a book in which he describes his experience of treating AIDS patients with ascorbate.[472] The book claimed that even people with fully developed, end stage AIDS can be helped with massive doses of ascorbate. He reported case studies of people recovering from full-blown AIDS and going back into an apparently stable remission. Since his book was published some time ago, we emailed Ian Brighthope to ask if he still held these positive opinions. His reply was rather poetic, but telling: "My position with respect to vitamin C is as strong as it has always been - vitamin C saturated individuals are healthy, generous, loving and peaceful - the solution to the world's ills!"

Brighthope's observations with AIDS support those of Cathcart, who has also reported on his extensive clinical experience of treating AIDS patients with ascorbate. Cathcart suggested vitamin C as a treatment for AIDS in 1983,[473] and published his first clinical report in 1984.[474,475] He reports that vitamin C can double the life expectancy of an AIDS sufferer, suggesting that ascorbate should be combined with other treatments for secondary infections. Many eminent physicians and nutritionists use ascorbate as the basis for their AIDS treatment, supporting Cathcart's approach.[494] Dr Robert Calapai suggests that ascorbate should form part of an aggressive but non-toxic therapy, to inhibit the replication of HIV. Dr Joan Priesley agrees that large doses of vitamin C should be a cornerstone of AIDS therapy. Nutritionist Dolores Perri uses infusions of up to 100 grams in her clinic. These doctors and others have reported significant benefits of ascorbate treatment.

Despite the fact that it was relatively easy to obtain funding for AIDS research and clinical reports suggested ascorbate might be effective, support was not provided for the study of vitamin C treatments. People diagnosed with AIDS often choose to take vitamin C, despite the lack of scientific follow up by the medical authorities. However, oral supplementation, while likely to be somewhat beneficial, is not going to reverse AIDS. Reports of successful treatment have used high dose, intravenous infusions of sodium ascorbate.

Ascorbate has unique effects in the treatment of AIDS.[474] In high doses, as expected, it acts as a powerful free radical scavenger and suppresses the HIV cell activation induced by oxidation. In AIDS and HIV infection, body levels of micronutrients are lower; this could lead to oxidative stress, indicating that antioxidant therapy could be beneficial.[476,477] Ascorbate slows the replication of HIV and the progression of the disease.[478] It also stimulates the undamaged elements of the immune system and provides a defence against secondary infections.

Studies have shown that ascorbate inhibits the HIV agent more effectively than AZT, the standard conventional drug.[479,480,481] Derivatives of ascorbate also have anti-HIV activity.[482] Vitamin C has been reported to inactivate the AIDS virus in the laboratory, while not having the damaging side effects on normal cells found with AIDS antivirals such as AZT.[483,484] The finding that an ascorbate derivative can selectively kill cells in which HIV is growing has reinforced this result.[485] The addition of other known antioxidants, such as glutathione and the dietary supplement NAC (N-acetyl-L-cysteine), amplified the antiviral effects. Vitamin C does not kill the virus directly but inhibits reverse transcriptase activity. Reverse transcriptase is an enzyme used by viruses as part of the process of replication. These findings suggest that vitamin C can stop, or greatly slow, the growth of the AIDS virus.

Given the history of vitamin C, it is inevitable that someone would find a reason for not using high doses of the vitamin in the treatment of cancer and AIDS. Eylar and colleagues reported that purified human T-cells deteriorate if maintained in culture in ascorbate for 18 hours.[486] Based on this single laboratory finding, Eylar suggested that dietary levels of ascorbate of up to a gram or so daily may be beneficial, but cautioned against using larger doses over a long period for the treatment of AIDS and cancer. The relevance of this observation is highly questionable, in view of clinical evidence supporting the safety of equivalent intravenous doses. Furthermore, cultured cells are often abnormal.

In 1980, Park reported that although several leukaemic cell cultures were sensitive to ascorbate at concentrations that could be attained in the body by injection, normal haemopoietic cells, which form the cells of the blood, were not suppressed.[550] Helgestad recently reported that vitamin C killed cells from a new malignant T-cell line from a boy with malignant lymphoma.[548] This indicates that extreme levels of ascorbate are more toxic to abnormal or cancerous cells than to normal tissue. Eylar's cell culture sensitivity is unlikely to apply to normal cells in the body.

To put Eylar's suggestion into context, we must remember that the side effects of conventional treatments for a person with AIDS or cancer are generally undesirable. In cancer, the alternatives are radiation and cytotoxic drugs, and for AIDS they are long-term anti-retrovirals. The absence of reported side effects with vitamin C, so far, indicates that the incidence is lower than with conventional treatment.

Robert Cathcart thinks that the problems of ascorbate treatment have been overstated. One conventional contraindication for high dose treatment is kidney disease, such as glomerulonephitis. Cathcart gives the example of one of his patients. "I remember the lovely school teacher who came to me several years ago. She said she bet that I did not remember her; she was the one whose mother came to me, fifteen years ago, when she was dying of glomerulonephitis in the hospital. I told the mother to sneak some ascorbic acid in massive doses to her in the hospital. A few weeks later she was out of the hospital, still somewhat oedematous (swollen) but recovering. I lost track of her for years but the daughter said she recovered completely with bowel tolerance doses and was perfectly well."[58]

Ebola

Ebola is an emerging viral disease, which competes with AIDS in terms of the fear it induces. Ebola is one of a number of newly discovered haemorrhagic fevers. A characteristic feature of haemorrhagic fevers is capillary fragility and bleeding. This disease has been described as turning the patient into a bag of soup, as the internal organs are liquefied and bleeding occurs from every orifice or break in the skin.

The symptoms of haemorrhagic fevers are so similar to severe scurvy, it can be difficult to distinguish between the two conditions. There was a widely reported outbreak of haemorrhagic fever in western Afghanistan in 2002. Later, it was found that the illness was not an

Ebola-like disease but scurvy, in an undernourished population.[487] Although acute scurvy is now an unusual disease, its symptoms are severe. In this case, it was mistaken for the effects of haemorrhagic fever, illustrating the similarity between the two illnesses.

In cases of scurvy produced by nutritional deficiency, the body is gradually depleted of ascorbate over a period of months. With haemorrhagic fever, most of the ascorbate in the body may be lost immediately, in fighting the virus. The virus quickly overcomes the body's antioxidant defences and the patient dies in a massive eruption of free radicals. The symptoms of Ebola and other horrific haemorrhagic fevers are entirely consistent with them being a result of severe induced scurvy. According to Cathcart, Ebola and the other haemorrhagic fevers produce a form of acute scurvy and the obvious treatment is massive intravenous infusions of sodium ascorbate. As no current treatment for this disease exists, there would appear to be nothing to lose from experimenting with vitamin C. A harmless treatment that might work would be worth trying on a patient who would otherwise almost certainly suffer a horrible death.

How does vitamin C fight infections?

We have described the action of vitamin C as a free radical scavenger; however, it has additional benefits in combating infectious disease. Bacteria and viruses do not cause disease by themselves: the body's immune system also has to fail. In any serious epidemic, some people fall ill and die whereas others are unaffected. Even with diseases as deadly as the haemorrhagic fevers, a few people are unharmed and antibodies to the organism may be found in their blood. A proportion of the population will be resistant to almost any disease. Some people even seem to be immune to AIDS.

The immune system

The immune system is a complicated arrangement of biological and biochemical mechanisms that has evolved to keep us healthy. There is a great deal of evidence that ascorbate is essential for the immune system to work effectively. Ascorbate stimulates the immune system and can help those with impaired immunity.[488,489,490,491,492] When a bacteria or virus enters the body, it is recognised by the antigen system and specific antibodies are produced in larger numbers. The antibodies selected for multiplication are those that can recognise and attack the invading bacteria or virus.

A high intake of vitamin C increases antibody levels in the blood. Ascorbate increases production of three types of antibody, called IgA, IgG and IgM or immunoglobulin A, G and M. These are the antibodies most associated with defence against disease. Notably, vitamin C does not increase the production of IgE, which is involved in allergic reactions and asthma. Vitamin C and other antioxidants may provide symptomatic relief for allergic conditions.[493] Cathcart argues that ascorbate provides exactly what we might have asked of it: increasing antibody response to disease but not allergies.

Invading bacteria may be attacked and destroyed by phagocytic white blood cells of the immune system. This process involves redox signalling and oxidation by free radicals; it is facilitated by vitamin C.[494] Low levels of ascorbate suppress the immune system. If vitamin C deficient guinea pigs are given skin grafts, the graft is tolerated. When the dose of vitamin C is increased, the skin grafts are rejected, indicating that the immune system is being activated. Experiments with megadoses of vitamin C show increased lymphocyte production, following antigenic stimulation.[495,496] In these studies, a dose of five grams per day doubled the rate of white blood cell production, 10 grams increased it by a factor of three, and 18 grams increased it to four times the base level.

White blood cells also require vitamin C for effective functioning, as described earlier. With higher concentrations of ascorbate, white blood cells become more active and can move towards infection or inflammation more quickly.[494,497,498,499] Large doses of vitamin C increase levels of interferons and provide a safe antiviral treatment.[500] Interferons are antiviral protein molecules, produced by cells during infection. Some have suggested that vitamin C levels in excess of 18 grams a day may be required for normal white blood cell functioning in cancer and infections,[2,501,502] but similar effects are likely to be achieved with lower doses, if blood levels are sustained.

Conclusions

The work on vitamin C as an anti-infective agent has a long history but, in scientific terms, it is incomplete. There are many clinical reports of its utility but no controlled clinical trials. The requirement for valid scientific studies to test massive doses of vitamin C in severe infections is clear. Until we have definitive studies in specific illnesses, very high dose vitamin C treatment is likely to remain an enigma.

The idea that the positive results obtained in treating with vitamin C are an example of placebo effect or other misleading factors takes us to the heart of the establishment position. For some reason, it strains the credulity of conventional medicine that anything as simple and innocuous as ascorbate could be an effective treatment for severe disease. They deem the possibility so low, that any simple experiments that apparently demonstrate an effect are assumed to be flawed.

When investigating unlikely phenomena, such as extrasensory perception or precognition, psychologists demand a higher level of proof. Normal science works at the 5% (or 1%) significance levels, which means that a result would not occur by chance alone more than one time in 20 (or 100). The basis of this approach is that it is more reasonable to accept the validity of an experimental result, if it were not likely to have occurred by chance. If you are investigating extrasensory perception, however, it is easier to believe the result was a one in 20 fluke than to accept an apparently magical property. To the conventional physician, the reported effects of vitamin C in infection border on the miraculous. This was especially true when no scientific mechanism was provided to account for the observations.

We asked a leading medical scientist about the results and observations of large doses of vitamin C in infection. He replied that he thought that they were explained by the placebo effect. He pointed out that he had successfully treated an alcoholic with placebo for over a year. Placebos are often almost as effective as standard prescription drugs. It is easy for physicians to convince themselves that what they are seeing is a real outcome when, in fact, it is only a placebo effect. While this interpretation has some general validity, it is clearly nonsense when applied to the reports on massive doses of vitamin C.

A placebo effect large enough to allow Klenner to cure polio, in 60 out of 60 patients, would be remarkable indeed. Remission from AIDS would also be unexpected. The placebo suggestion lacks validity, because no known drugs or treatments, placebo or otherwise, have an effect of this magnitude. Since the scale of the effect is so large, the alternative suggestion, that the effectiveness of vitamin C can be treated as pseudo-science, also lacks credibility. The appropriate response would be replication or refutation, by scientific studies. We note that negative results would be most welcome to the pharmaceutical industry, whereas positive results might be financially catastrophic.

The hostility of the medical establishment to the idea that vitamin C may be helpful for trivial infections, such as the common cold, is bizarre. Many patients will not tell their doctors what supplements they are taking, because of a perception or experience of this ignorant attitude.[503] Rather than increasing the standing of the physician, this level of arrogance may lower the perceived validity of the doctor's advice, driving the patient towards alternative practitioners.

With more severe viral diseases and antibiotic resistant bacterial infections, where there is no known effective treatment, vitamin C offers hope. There is no reason not to try Cathcart's approach and give large intravenous infusions of sodium ascorbate. Scientific evidence and substantial clinical experience suggests a powerful effect, from a safe treatment. In the absence of controlled clinical trials, we are forced to conclude that to withhold treatment with massive doses of intravenous ascorbate in deadly incurable diseases, such as rabies, SARS or Ebola, is unethical.

Cancer

"Dose, Timing, Duration." Dr Thomas Levy

Cancer is not a single illness but a large number of different ones, involving abnormal cell division and growth. The primary feature of cancer is cell proliferation. A single cancer cell can produce billions of offspring by cell division, forming a large tumour. When growing rapidly, cancer cells can move and invade the surrounding tissues. During the infiltration process, the cancer has to overcome factors that normally limit tissue growth. A tumour may produce enzymes, which break down the surrounding tissues and make it easier for the tumour cells to spread. This process is active and aggressive; invasive tumours are even able to destroy bone.

Cancer patients often lose weight rapidly, as the tumour takes in large amounts of nutrients to feed its disturbed metabolism. Where oxygen is in short supply, cancers can switch to a primitive form of metabolism, similar to fermentation, which requires less oxygen. A characteristic of cancers is that they are able to seize large amounts of nutrients and oxygen. They do this by releasing free radicals and other substances, to promote the development of blood vessels. One approach to cancer treatment, pioneered by Dr Judah Folkman,[504] is to block the formation of new blood vessels required by the tumour; this means that tumours cannot grow and so remain small. In theory, this treatment has great potential but its application in humans is still under investigation.

Types of cancer

Cancer is described in terms of the original site of the tumour, for example, ovarian cancer begins in the ovaries and bowel cancer starts in the large intestine. We will concentrate mainly on tumours or mass-forming cancers. However, what we discuss also has implications for other forms of cancer, such as leukaemia. Tumours are cancerous masses that come in two forms: benign and malignant. Benign tumours are usually slow growing, non-invasive and relatively safe. A typical example is a lipoma, which is a tumour of fatty tissue, sometimes found under the skin. Lipomas are often encapsulated, well defined and distinct from the surrounding tissues. By contrast, a malignant tumour is an expanding mass of tissue, which normally does not have a well-defined border or surrounding capsule. Malignant tumours invade neighbouring tissues, often infiltrating them and preventing the encapsulation found in benign tumours. A malignant tumour feels more attached and part of its surroundings than a clearly distinct benign lump.

Tumours are classified into two main groups, carcinomas and sarcomas. The rest belong to a group of miscellaneous others. Carcinomas originate from epithelial and endothelial tissues. Epithelial tissues form the coverings of the external and internal surfaces of the body, and include skin and the linings of the mouth and gut. Endothelial tissues form the inner lining of the blood and lymph vessels, and the heart. Carcinomas are relatively common; one reason for this is that the epithelial tissue is generally the first to encounter cancer-causing agents.

Sarcomas arise from connective and non-epithelial tissues, such as bone, cartilage, fat or muscle; they form less than five percent of common tumours. The distinction between tumour types is somewhat arbitrary, in the sense that cancers lose the characteristic features of the cells from which they were originally formed. For example, kidney cells are different from other cells in the body: the biological process that leads to this specialisation is called differentiation. However, kidney cells that proliferate and form a tumour are damaged and the resultant cancer cells have little in common with the original kidney. Cancer cells are often completely undifferentiated, having lost the character of the tissue from which they were derived.

Early Studies

In 1940, Voght published a study on the use vitamin C in the treatment of leukaemia, suggesting that cancer patients often have a substantial deficit.[505] The deficit was corrected with intravenous ascorbate. The results were suggestive of a therapeutic effect, although dose levels designed to overcome a deficit would be unlikely to lead to the high tissue levels required for a treatment. In 1954, McCormick proposed a relationship between cancer and vitamin C.[506,507] His hypothesis was that vitamin C protects against cancer by increasing collagen synthesis. McCormick had noticed that cancerous tissues had a remarkably similar appearance to tissues from people with scurvy. He suggested that cancer was a collagen disease, related to a lack of ascorbate.

Nowadays, cancer is understood to be a disease of cells, rather than the collagen-containing connective tissue that supports and separates tissues. Nevertheless, when a cancer grows, the diseased cells invade normal tissue. It is possible that strengthening the connective tissue would make it more difficult for the cancer to grow and spread.

McCormick's observations were instrumental in leading Linus Pauling and Ewan Cameron to study the benefits of vitamin C in cancer patients, over a number of years.[501,502,508,509,510,511,512,513,514] Their studies include reports of complete cures.[515]

In the 1970s, Irwin Stone proposed a definite link between vitamin C deficiency and cancer.[516,517] He pointed out that vitamin C had been used as a treatment for cancer of white blood cells by Dr Greer, in 1954. Greer had treated a patient suffering from myelogenous leukaemia, using doses of 24 to 42 grams, and had subsequently observed a complete remission of the disease.[518] Greer indicated that the remission was a result of the vitamin C. He stopped the treatment twice and the patient deteriorated both times. When the vitamin C was discontinued, the patient felt ill, his temperature went up and other symptoms returned. On each of the two occasions when ascorbate was resumed, the patient's temperature returned to normal within six hours, his malaise disappeared, and the disease apparently went back into remission. Despite searching the literature published over the following 20 years, Irwin Stone was unable to find a replication experiment or any follow-up to this exciting clinical report.

Cytotoxic agents

A cytotoxic drug is a chemical that kills cancer cells directly. It is characteristic of cytotoxic drugs that they are poisonous to both the diseased tissue and to normal body cells. The primary aim of research into this form of anticancer drug is to find a substance that is more poisonous to cancer cells than to normal tissues. An ideal cytotoxic drug would kill the cancer cells at a lower dose than it destroys normal cells. In practice, both cytotoxic drugs and radiation therapy kill rapidly dividing cells more effectively than they do slowly growing cells. The appalling side effects of these drugs are a result of cell death in quickly growing normal cells, such as those in the bone marrow, the lining of the gut and the hair follicles. Such drugs are often most effective against rapidly dividing, aggressive cancers. Unfortunately, most of the big killers are slow growing cancers, which are more resistant to the effects of cytotoxic drugs.

Cancerous tissue is dangerous because more cells are produced than die, causing it to grow. Most tumours contain more than one type of cancer cell. This diversity occurs because cancer cells are damaged and their cell division is prone to errors, resulting in alterations to the descendant cells. Consequently, a typical tumour could contain cells that are dividing rapidly, slowly, not at all, or are dying. Cancer cells are subject to a form of evolutionary pressure that selects the most rapidly growing and robust cells.

Tumours are varied populations of cells, which evolve to resist the effects of treatment. When a cytotoxic drug, or x-ray, kills the more aggressive cells, the hope is that the balance will be switched to more cells dying than growing, and the tumour will therefore regress. With luck, the chemotherapy will eradicate the tumour or, at least, additional time will be added to the patient's lifespan. Often when the drug is discontinued, the remaining cells respond to the evolutionary pressure and fast-growing cells proliferate. Like bacteria becoming resistant to antibiotics, this new population of cancer cells may be more resistant to the treatment.

Drug testing

People would probably expect a new cancer drug to go through several levels of testing, to show it is more effective, safer or cheaper than alternatives, but this is not necessarily the case. Italian pharmacologists Silvio Garattini and Vittorio Bertele studied cancer drugs entering the European market over the period 1995 to 2000.[519] They reported that new drugs offered few important advantages over existing drugs, but could cost up to 350 times more. New drugs were being introduced merely to increase drug company profits. They noted that treatment of cancer with cytotoxic drugs offers only a small chance of cure, but such drugs were used as a treatment of last resort, especially when the cancer had spread throughout the body. These new drugs were often introduced as second or third line treatments for rare cancers, so that they would be accepted and get past the regulatory hurdles more quickly. Garattini and Bertele recommended that new drugs should be required to show substantial clinical benefits or cost reductions before being accepted.

Host resistance and spontaneous recovery

Most people do not get cancer for long periods of their lives; one reason for this is "host resistance". Immunological and other mechanisms prevent the cancer from getting started. However, the incidence of cancer increases dramatically when a person's immune system is damaged or suppressed. Detailed knowledge about host resistance is inadequate, but we do know that it is likely to be involved in the success of normal clinical treatments.

Host resistance can be examined scientifically. People have inbuilt mechanisms, which either prevent cancers occurring or eliminate them at an early stage. These come into play during conventional treatments, such as surgery. Successful surgery generally leaves millions of cancer cells in the bloodstream, but the patient still recovers. When a large tumour is removed, the remaining cells can often be dealt with by the immune system or by encapsulation, so they do not go on to form another large mass.

Surgery is not generally recommended when a tumour has metastasised, which means that cells have broken away and formed discernable lumps in other parts of the body. If one additional site has become large enough to be found by a clinical test, such as an x-ray, the cells have probably spread to many other parts of the body. In these cases, removal of the original tumour is unlikely to lead to a cure. Sometimes, however, if the original large tumour is removed, small metastatic cancers shrink away, by mechanisms that are not completely understood.

In rare cases, the biological mechanisms that help eliminate the final few cancer cells may produce an apparently complete cure. These rare miracles, known as spontaneous recovery, are important, because if the reasons why these patients recover were understood, the information might be used to help the treatment of others. Unfortunately, with spontaneous remission, we are at about the same level of science as stamp collecting, as by definition, we cannot repeat the process. Since replication is a critical feature of science, the study of spontaneous regression remains on the fringes. We can only guess at the mechanisms that result in a person with terminal cancer suddenly becoming healthy. One frequent speculation is that the patient's immune system is stimulated into action and has killed the cancer.

Recognised spontaneous remission is rare but, over the years, hundreds of cases have been documented.[520,521] By 1966, an estimated 200-300 accepted remissions had been reported, but the actual number is likely to be much larger. Some cancers are likely to self-heal before they are diagnosed and many claimed medical cures could be the result of spontaneous remission. Such remissions are a dramatic example of how, in the right circumstances, the body can deal with cancer.

Vitamin C as a treatment for cancer

"The present study shows that ascorbate (vitamin C) is highly toxic or lethal to Ehrlich ascites carcinoma cells in vitro." Dean Burk (1969)

The importance of finding a safe substance that can kill cancer cells is difficult to overemphasise. Conventional chemotherapy drugs that kill tumour cells are poisons. A safe agent that killed a wide range of cancers is the holy grail of cancer research. It would be the cancer equivalent of discovering penicillin.

Treatment of cancer: clinical studies

The idea that vitamin C can be used to treat cancer was popularised in 1976, by Linus Pauling and Ewan Cameron, a Scottish surgeon.[502,511] Cameron was a respected and established cancer specialist; his work was solid and conventional.[3] He believed that the most important factor in recovery from cancer was the person's biological response. In 1972, he published a paper suggesting that vitamin C might slow cancer growth by increasing collagen production and inhibiting the enzyme hyaluronidase, which is released by cancers to dissolve surrounding tissue.[508]

Cameron himself was initially sceptical that vitamin C could work against such a baffling and complex disease as cancer. Despite his reluctance, he was convinced that his patients had nothing to lose. They were in the terminal stages of their illnesses and, even if the treatment did no good, it would at least do no harm. He started giving 10 grams per day of vitamin C to patients with untreatable cancer and soon became convinced that most benefited from it.

Cameron told his colleagues but found that they were unable to accept that such an innocuous substance could possibly work. The contrast with conventional cancer treatments, such as radiation, major surgery and toxic chemotherapy, was too great. In some cases, the usual treatments seemed worse than the disease itself. Cameron's ascorbate approach was attacked with arguments that ranged from differences in treatment methods through to the philosophy of medicine. Cameron, himself incredulous at the start of the experiments, had anticipated this negative response.

Cameron and his collaborators provided clear documentation of their work with vitamin C,[502,508,509,510,511,512,513,514,515] over a number of years.

To begin with, Cameron and Pauling published case reports of 50 patients, who were given ascorbate injections along with oral supplements. They then expanded the number of patients to 100, reporting that cancer patients treated with vitamin C had survived three to four times longer than had control patients. The control group was based on the records of 1,000 patients who had received no vitamin C, in a study centred at the Vale of Leven Hospital, near Loch Lomond, Scotland. The controls were matched (10 to 1) with the ascorbate treated patients, with respect to age, sex, type of cancer and clinical status of "untreatability". They were treated in the same hospital, by the same staff and were managed identically, except for the supplemental ascorbate. An outside doctor, who had no knowledge of the treated patients, examined the control patient records and recorded their survival times, defined as the time in days between the end of conventional treatment and the date of death.

Traditionalists argued that the control group selection in Cameron's studies might have been biased. They suggested that the vitamin C and control groups were not properly matched and might have had a different severity of disease. They claimed that taking control patient details from medical records might also introduce bias. A bias in the selection, as suggested by the detractors, could theoretically explain the positive results, but would need to be extreme. This criticism was rather harsh for a preliminary study. Both Pauling and Cameron would have preferred to have the resources to conduct a more rigorous trial, but funding was repeatedly denied.

Robert Cathcart has made a more important objection to the Vale of Leven studies, namely, that Pauling and Cameron had not given a large enough dose. Ten grams is at the low end of clinical doses and would be insufficient to provide relief from a bad cold. Animals that manufacture their own vitamin C get cancer. Since these animals often have high tissue levels of ascorbate, this sets a lower limit on the effective dose. Pauling and Cameron's positive findings were therefore unexpected, but might be explained by recalling that the ascorbate was administered intravenously, which leads to higher tissue levels.

A prospective cancer study was conducted from 1978 to 1982 in the same region of Scotland, with 294 patients treated with ascorbate and 1532 controls.[502] Patients were not randomised and received vitamin C or palliative care according to which doctor admitted them. Patients receiving vitamin C had an average survival period of 343 days against only 180 days for the controls. Furthermore, the supplemented patients appeared to have an improved quality of life, with reduced pain. These results were suggestive of a powerful effect but were also flawed.

The first problem is that the controls were not exposed to exactly the same conditions as the treated patients. The selection of patients was not random and was therefore potentially subject to bias. The different survival figures could simply reflect unconscious preference by the physician in selection or in initial diagnosis. In addition, there were no placebo controls, as Cameron, being convinced of the efficacy of the treatment, was ethically unable to deny vitamin C treatment to dying patients. A cancer drug could not be used as a placebo, as the patient would need to take it for the rest of their lives. A double-blind clinical trial would have been revealing, but it would need to be carried out by researchers who had yet to be convinced of the merits of vitamin C.

In 1979, Morishige and Murata published a study confirming the results obtained by Cameron.[522] They studied 99 patients, of whom 44 subjects received four grams of vitamin C or less per day and 55 patients received five grams or more. The death rate for the higher dose ascorbate patients was only one third that of the patients receiving the lower doses. Patients receiving the low dose lived an average of 43 days. Those receiving five to nine grams lived 275 days and those receiving 10 to 15 grams lived an average of 278 days. Rather surprisingly, patients receiving the highest doses, 30 to 60 grams, lived an average of 129 days. This is three times longer than the lowest dose patients but only half as long as the 5-15 gram groups. The Japanese researchers' explanation for this finding was that the highest doses were given to the patients with the most advanced disease.

This Japanese experiment replicated Cameron's study, as the ascorbate treated patients lived far longer. Unfortunately, this study was also not a double-blind clinical trial, so again the control for placebo effects and experimenter bias was not complete. However, the groups with different dose levels acted as internal controls. Unless there was substantial bias in selection between the treatment groups, this replication showed that vitamin C prolonged life in patients with cancer. However, the medical community found the results of the Japanese study questionable and did not accept its validity, because of the lack of double-blind randomised controls.[523] Considering the lack of progress in cancer treatment and the toxicity of current chemotherapy, this stance might be considered dogmatic.

The Mayo Clinic study

In response to what they considered ridiculous claims about the benefits of vitamin C, the prestigious Mayo Clinic decided to conduct its own study.[524] It has since been suggested that their real aim was to quash the controversy surrounding vitamin C and cancer.[525] While the Pauling and Cameron study lacked rigour, the Mayo Clinic study fuelled even more controversy.[526,527] Charles Moertel, the principal Mayo clinic investigator, chose not to discuss his experimental design and analysis of results with Cameron and Pauling. He was within his rights to do this and a supporting argument would be that he wished the study to be independent. This refusal to collaborate carries with it the responsibility to ensure a rigorous study. In retrospect, it is unfortunate that collaboration was denied, as Pauling might have pointed out the limitations of the Mayo Clinic's experimental design.

On the plus side, the Moertel study was prospective and double blind, with randomised controls. Ten grams of vitamin C were given *orally* to the subjects and a placebo to the controls. Care was taken that the 63 controls matched the 60 supplemented subjects. About a quarter of all subjects had an improved appetite, and a larger proportion had increased physical activity with some nausea and vomiting. Both groups survived for about the same length of time, seven weeks. The study concluded that there was no benefit from vitamin C.

Linus Pauling responded that Moertel's patients had previously received chemotherapy and this would have damaged their immune systems. Vitamin C would be expected to be less effective on such patients, as one of its proposed mechanisms was stimulation of an immune response to the tumour. The controls could also have been taking vitamin C supplements independently and the researchers might have been gullible in this respect. A subject in an experiment to test whether or not vitamin C was a cure for cancer could realise they had a 50-50 chance of being in the control group and might self supplement, to be on the safe side.

Further arguments included the suggestion that Moertel's patient sample was biased, in that his patients did not represent the large majority of cancers. Moertel then conducted another study to answer these objections.[528] Once again, there was no collaboration with Pauling's research group. In this placebo-controlled study, samples of urine were taken from a few patients in both control and supplement groups; these were checked for vitamin C concentration. Proponents of vitamin C continued with their objections, including the poor quality of the check on self-supplementation and the fact that patients were supplemented only for a limited time, instead of continuing for life.

There were clear problems with the Moertel studies. Indeed, Pauling implied that they were an example of scientific fraud,[2,502] and that the claimed results had been reported in a way that was deliberately misleading.[3] The medical establishment, however, bought into the Moertel study, perhaps thinking it inconceivable that the prestigious Mayo Clinic would be involved in such skulduggery. It therefore became accepted medical dogma that vitamin C had been tested and found wanting. The New England Journal of Medicine delayed publication of Linus Pauling's response to the Moertel study by about two years, until Pauling's lawyer wrote to the journal. The Journal responded that they would not publish anything further on the matter, unless it was a double-blind clinical trial. This delay helped establish the idea that vitamin C was of little benefit in the treatment of cancer.

Rather than being forced to repeat our discussion of the limitations of double-blind clinical trials, we merely quote the recent satire on the use of parachutes in the British Medical Journal by Smith and Pell.[529] They state that after a search of the medical and scientific literature, they could find no studies of the effectiveness of parachutes on preventing injury caused by falling from great height. "As with many interventions intended to prevent ill health, the effectiveness of parachutes has not been subjected to rigorous evaluation by using randomised controlled trials. Advocates of evidence-based medicine have criticised the adoption of interventions evaluated by using only observational data. We think that everyone might benefit if the most radical protagonists of evidence-based medicine organised and participated in a double-blind, randomised, placebo-controlled, crossover trial of the parachute."

There is more to the analysis of experimental results than double-blind clinical trials. Where a double-blind experiment with placebo controls cannot be done for ethical or other reasons, other methods of analysis do exist. When faced with the unacceptability of non-placebo controlled studies and the ethical problems of a doctor who believed in the value of a treatment, Linus Pauling looked at alternative methods of analysis. He came up with a method based on the Hardin Jones principle.[533,534]

Hardin Jones was a critic of cancer research methods and believed that most cancer research was incorrectly performed. By analysing cancer statistics, he discovered that the expected death rate is constant for a given type of cancer; this is the Hardin Jones principle. That is, a cancer patient has the same probability of death on any day. This idea seems strange at first but the mathematics is valid. If this assumption is made, then a comparison of factors affecting the expectation of longevity can be carried out more easily. In principle, this method could provide an analysis suitable for less controlled studies. This approach could reduce bias in clinical trials. However, it does not remove the necessity for random selection into treatment groups. It merely replaces one assumption (that the selection into groups was unbiased) with another (that the groups have similar life expectancies).

Designing a good experiment is difficult and requires great expertise. It has been pointed out that Moertel's Mayo Clinic studies do not fulfil the requirements of the Hardin Jones method of analysis as, for example, the treatment was for a short time interval rather than continuous.[530] None of the patients treated with vitamin C died during the treatment phase of the study (median time 2.5 months). Some died when vitamin C was discontinued, others when they were given chemotherapy after concluding participation in the study.

Despite repeated requests, the Mayo Clinic refused to release their raw data for analysis by other scientists. This refusal to release the raw data suggests a lack of confidence in their published analysis. The available data suggest that surviving patients in the Mayo Clinic had their treatment changed in such a way as to increase the death rate. A possible explanation of this might be that chemotherapy reduced the life expectancy of the longest surviving patients.

It appears that Linus Pauling did not point out to Charles Moertel the crucial difference between intravenous and oral vitamin C. There is no reason why he should have needed to: it is basic beginner's pharmacology. An intravenously injected dose goes straight into the blood plasma and its full effect is felt immediately. Even to someone with little more than a basic understanding of biology and the action of drugs, it should be obvious that injections have a more direct action than oral administration. When large doses of sodium ascorbate are given intravenously, blood concentrations can reach levels at least 10 times higher than an equivalent oral dose. In 2002, Gonzalez and colleagues reported that intravenous ascorbate gives a more consistent result in cancer patients, since the blood levels attained are higher.[541] Given this knowledge, Moertel's switch to oral doses would clearly have biased the results.

Serendipity?

Sebastian Padayatty and Mark Levine also concluded that intravenous ascorbate is more effective in cancer treatment, as it enables higher blood levels to be reached.[531] They thought this discrepancy may be one cause of the differences between the Pauling and Mayo Clinic studies. Surprisingly, they suggest that Pauling and Cameron might have "serendipitously" stumbled across an effective treatment. We find their implication that Pauling and Cameron did not understand the difference between intravenous and oral administration bewildering. It is also demonstrably inaccurate, as Pauling wrote about the greater effectiveness of injected doses. He stated clearly that intravenous doses are more effective than oral doses and explained the reasons for the difference.[2] More recent results have led Padayatty and Levine to propose that the time is right for a re-evaluation and trials of vitamin C in cancer.

Padayatty and Levine suggest that Linus Pauling and the physicians he worked with were not aware of the cytotoxic action of vitamin C. This is extremely unlikely, as a study in 1969 by Dean Burk showed direct killing of cancer cells by ascorbate.[546] The author stated that, "In our view, the future of effective cancer chemotherapy will not rest on the use of host-toxic compounds now so widely employed, but upon virtually host-non-toxic compounds, that are lethal to cancer cells, of which ascorbate represents an excellent prototype." This finding remains common knowledge among vitamin C researchers; it is covered in Irwin Stone's book, which has a foreword by Linus Pauling.[437] Cameron and Pauling initially concentrated on presenting the case for vitamin C strengthening the surrounding tissues and encapsulating tumours; this emphasis is understandable. Since the idea that vitamin C could be used to treat cancer was greeted with hostility and aggression from the medical establishment, it would have been natural to limit the discussion to areas that were less controversial.

Consequences of the Mayo study

The Mayo Clinic study has discouraged many people with cancer from taking vitamin C, despite the fact that other research reported large benefits. If Moertel were correct, we could justify rejecting use of the vitamin. For this to be the case, the experimenter bias in the positive studies would have to have been substantial. If Moertel were wrong, then mistakenly rejecting the positive vitamin C findings would cost tens of thousands of lives. The Moertel studies would need to be extremely reliable to risk so many lives. However, they are unreliable, weak and biased.

We are not going to enter into the detailed arguments about the Cameron and Moertel studies. The way each was conducted is open to criticism and both were subject to more than heated debate.

Unfortunately, Moertel chose to limit Pauling's access to the data and to avoid direct discussion. Part of the reason for this may be that Moertel and his researchers were scared of Pauling's combative approach. Great genius comes in many forms. Albert Einstein had a saint-like reputation; if you suggested he had made an error, he would reputedly have little ego in considering your objections. Linus Pauling's reputation was more that of a scientific gunslinger, with a career-long ability to defend his position against all comers. In the 1980s, Linus Pauling was more than capable of dealing with the likes of Charles Moertel and the Mayo Clinic experts.

Some believe that the dispute between Pauling and the Mayo Clinic could not be resolved within the normal confines of science.[532] They argue that medicine is not science and the dispute was as much political as scientific; the adversaries were defending positions of power and influence. There is some merit in this idea but it is misguided. Like Thomas Kuhn, with his ideas of science progressing by way of paradigm shifts and scientific revolutions, it tries to turn science into sociology. Perhaps followers of such ideas have forgotten the tales of King Canute or The Emperor who had no Clothes. Despite what people believe, the biochemistry remains the same.

Positive results continue

Despite Moertel's rejection of the idea, reports of the efficacy of vitamin C as a treatment for cancer have continued. Perhaps the most notable are from Dr Abram Hoffer, who was the first to use a double-blind clinical trial in psychiatry. His results confirm those of Cameron, Morishige and Murata, in that they show greatly increased survival times.[533,534] Hoffer's book provides a wealth of information on his own studies, including both the analysis of results and extensive case study presentations.[3] Like Cameron, Hoffer was soon convinced that the treatment worked, and became unable to perform placebo controlled clinical trials, because of ethical considerations.

One of Hoffer's subjects was a female patient, who had read Norman Cousins' account of treating his illness with vitamin C,[4] and started taking 10 grams each day. This woman had pancreatic cancer and her doctors had given her only a few months to live. She was referred to Hoffer, who had an interest in nutritional supplements and disease. Hoffer increased the dose to 40 grams and, after six months, the woman's cancer was reported as having disappeared. The response intrigued Hoffer, who had previously treated cancer patients successfully with high doses of niacin and vitamin C.

With the accumulation of scientific evidence, the establishment view is beginning to change as their former position is becoming less tenable. In an important step, Riordan and others have confirmed that vitamin C is toxic to cancer cells, in the laboratory,[535] in animal studies,[536] and in the body.[537] It does not take a genius to understand that this would explain the positive results in the studies of vitamin C and cancer. Even a biased establishment can see the implications. Riordan has published case studies indicating that vitamin C is an effective cancer treatment,[537,538] and has also published a detailed mechanism of action.[539,540,541] Derivatives of vitamin C have been shown to have anti-tumour actions,[542,543] and a recent paper proposes encapsulation of vitamin C in micro-particles for use as an anticancer treatment.[544]

Case study

A typical case study, provided by Riordan, describes a 70-year-old white male who, in December 1985, was found to have a tumour of the right kidney. When the kidney was surgically removed, it was shown to contain renal cell carcinoma. Three months after surgery, the patient had a standard x-ray and a CT scan. He was found to have "multiple pulmonary lesions and lesions in several areas of his liver, which were abnormal, and some spread to the lymph glands". These secondary tumours implied that the patient's life expectancy was short.

In March 1986, the patient was seen in Riordan's clinic. He chose intravenous ascorbate rather than conventional chemotherapy. He was given 30 grams of vitamin C intravenously, twice a week for seven months. The same dose was then given once a week for eight months, followed by 15 grams a week for a further six months. There were no apparent toxic effects and his blood and urine samples were normal.

Only six weeks after starting treatment, the patient felt well and the oncologist's report stated, "His exam is totally normal. His chest x-ray shows a dramatic improvement in pulmonary nodules compared to six weeks ago. The periaortic lymphadenopathy (enlargement of the lymph nodes surrounding the aorta) is completely resolved ... either he has had a viral infection with pulmonary lesions with lymphadenopathy that has resolved or (2) he really did have recurrent kidney cancer which is responding to your vitamin C therapy."

Now we fast-forward 10 years, to July 1996. The patient's current oncology report states: "there is no evidence of progressive cancer. He looks well ... chest x-ray today is totally normal. The pulmonary nodules are completely gone. There is no evidence of lung metastasis, liver metastasis or lymph node metastasis today, whatsoever." The patient died in 1997, aged 82 and apparently free of cancer, 12 years after his diagnosis.

Vitamin C kills cancer cells

As early as 1969, vitamin C was shown to kill cancer cells directly.[546] Klenner suggested the use of massive dose intravenous ascorbate as a treatment for cancer in 1971, which was well before the Cameron and Pauling and the Mayo Clinic studies.

It is difficult to extrapolate from test tube studies to an anti-cancer effect in the body. To show a useful cytotoxic effect, it is necessary to demonstrate the effect in the test tube, show that the effect can also occur in the body and, finally, obtain clinical results indicating a measurable reduction of tumour size and improvement in the patient's clinical condition. This has now been done. What remains is for the experiments to be replicated and suitable double-blind clinical trials to be performed.

Recent research by scientists interested in vitamin C and cancer has concentrated on large intravenous doses. Riordan studied high dose ascorbate as a cancer treatment. At first, doses of only 15 grams were given, once or twice each week. These doses reduced pain and improved patients' sense of well-being, as well as prolonging life longer than expected. More than a decade ago, Riordan started using doses of 30 grams of intravenous vitamin C, given twice weekly. Later he reported a case in which 100 grams of ascorbate, given intravenously once or twice a week, resolved bone metastases in a patient with breast cancer. Observational evidence can sometimes be sufficiently strong as to be persuasive. When patients with metastatic cancer recover and the clinical report is one of many, the evidence becomes compelling.

More recently, Riordan confirmed that ascorbate could do more than alter the tumour growth rate; it could also kill cancer cells. Moreover, he showed the vitamin was toxic to many different types of cancer cell. Crucially, Riordan found that ascorbate kills tumour cells at a dose that appears safe to normal cells. This finding is supported by several independent studies, which have also shown vitamin C to be toxic to tumour cells in both tissue studies and animal models.

A new treatment?

Showing that a chemical kills tumour cells, at lower concentrations than it harms human cells, is only the first step in getting a new treatment. In most cases, a new therapy has to go through detailed toxicity testing and be capable of delivery to the target tumours. With vitamin C, many of these issues simply do not apply, as ascorbate is a normal and essential part of the body. It is necessary, however, to show that a suitably high concentration of vitamin C can reach the cancer.

Riordan has tested samples of human serum from patients receiving intravenous ascorbate. His measurements confirm that the levels obtained are cytotoxic to tumour cells in experimental studies. A concentration of 400mg per litre was reported as sufficient to kill tumour cells, although this was from experiments using sparsely populated layers of cells, in a standard tissue culture. The levels of vitamin C measured in blood plasma following intravenous injection were high enough to be toxic to cancer cells.

In his experiments, Riordan demonstrated that samples of blood plasma, taken following ascorbate injections, killed cancer cells in tissue culture. To model the conditions in the human body more closely, Riordan checked the findings in experiments with a three-dimensional tumour model. In these later experiments, he used human blood serum as a growth medium. Using this more elaborate model, the effective concentration to kill most types of tumour cell was increased to four grams per litre. By measuring the blood levels of ascorbate levels during and after intravenous infusions, Riordan's group confirmed that a 60gram dose produced peak plasma concentrations that were more than double those required for killing cancer cells.

Riordan and others have suggested a way that ascorbate could kill cancer cells.[545] The test tube results on the killing of cancer cells by vitamin C are somewhat dependent on the culture medium. They proposed that ascorbate does not kill cells directly, but acts by producing hydrogen peroxide. Riordan found that the concentration of the enzyme catalase, which breaks down hydrogen peroxide, is up to one hundred times greater in normal cells than in tumour cells. Agents that cause hydrogen peroxide to be generated should therefore be toxic to cancer cells, while being safe for normal tissues. In this way, ascorbate at a sufficient dose should kill tumour cells, with no toxic effects to normal body cells.

Both ascorbate and metabolites of ascorbate have shown anti-tumour activity in isolated tissues. Research by Dean Burk at the US National Cancer Institute first indicated that ascorbate was highly toxic to carcinoma cells.[546] Notably, the toxicity increased if a catalase inhibitor was present. Bram reported that vitamin C was preferentially toxic to experimental cell lines of malignant melanoma (skin cancer).[547] They found that copper increased the toxicity to the melanoma cells. The presence of copper can cause ascorbate to act as a pro-oxidant rather than an antioxidant. This can lead to greater production of hydrogen peroxide. Helgestad reported that a new malignant lymphoma T-cell line was sensitive to ascorbate in culture, at concentrations that were attainable in human blood.[548] Additional research indicated that several leukaemia cell cultures were sensitive to vitamin C concentrations that could be achieved in the human body, while normal blood forming cells were not suppressed.[549,550,551,552] Research into the action of vitamin C and selenium on gastric cancer also led to the suggestion that the combination may be a useful treatment.[553]

Vitamin K helps vitamin C

The addition of vitamin K increases the anticancer properties of ascorbate. Vitamin C, combined with a much smaller amount of vitamin K3, will kill cancer cells.[554,555,556,557,558] It has recently been proposed that vitamins C and K3 could be used for treatment of cancer.[559,560] Vitamin K is a fat-soluble vitamin, involved in blood clotting. These vitamins work together to provide anti-tumour activity. They kill tumour cells preferentially, by an unusual process called autoschizis.[561,562,563] This is a novel type of cell death, involving substantial membrane damage. The cell bleeds cytoplasm, the fluid it contains, leaving behind the sub-cellular particles and structures. Cells rapidly become smaller and surround a shrunken nucleus with a narrow ring of cytoplasm, containing the internal particles or organelles. The mitochondria also shrink but the cell does not appear to die from lack of energy. This process depends on a reduction in DNA synthesis and oxidation.[564] Moreover, addition of the enzyme catalase can prevent the cell death.[565]

Experiments on mice with tumours indicate that treatment with oral vitamin C and vitamin K3 increases their life span; tumours in the treated mice grow more slowly. Human prostate tumours implanted into mice are selectively sensitive to these vitamins.[566] Mice are not the ideal animals for vitamin C studies, however, as they do not need it in the diet. The effect may be greater in vitamin C dependent species, such as man or guinea pigs.

Studies on human bladder cancer cells show that they are sensitive to the action of vitamins C and K3.[567] These vitamins can also increase the effects of standard anti-cancer chemotherapeutic drugs.[568,569] Laboratory studies, by a Chinese research group, confirm that vitamins C and K3 kill a range of cancer cells.[570] In these experiments, two types of human cancer were more sensitive to the vitamins than were normal cells. A Japanese group has found that the cytotoxic activity of both vitamin C and K3 can be enhanced by the plant extract lignin F.[571]

Lipoic acid

Lipoic acid increases the power of ascorbate to destroy cancer cells. Riordan has tested the potency of an ascorbate and lipoic acid mixture for killing cancer.[572] He used a standard solid tumour model and clinically achievable concentrations. The addition of lipoic acid increased the anti-cancer effect; the concentration of vitamin C at which half the cancer cells were killed was reduced from 800 mg/dl to only 120 mg/dl. That is, only about one seventh the amount of ascorbate was required to kill cancer cells when lipoic acid was included. The treatment was effective against both rapidly growing and non-proliferating cells. This combination may be more effective than existing drugs.

Preventing cancer growth

Further confirmation that vitamin C could be an effective cancer treatment suggests that ascorbate prevents cancer cells from growing. Ki Won Lee and colleagues, from the University of Seoul, have described a mechanism by which vitamin C stops cancer cells proliferating.[573,574] Normal cells stop growing when they receive signals from neighbouring cells; these signals are induced by hydrogen peroxide. Cancers can continue spreading, because they do not respond to the signals that inhibit growth in normal cells. Ki Won Lee showed that ascorbate enabled cancer cells to receive and act on the message to stop growth.

Dangers with intravenous ascorbate

Occasionally, it has been reported that high dose intravenous ascorbate can destroy a tumour too quickly. One case study recorded a death, following only 10 grams of intravenous ascorbate. In this case, the vitamin C killed the tumour cells, producing necrosis and haemorrhage that killed the patient. The treatment was successful in destroying the cancer, but the patient died. These clinical observations reinforce the obvious need for medical supervision. However, on the plus side, these effects demonstrate that vitamin C can certainly destroy tumours within the body.

Conclusions

We have described evidence that vitamin C can kill cancer. Indeed, the properties of ascorbate reflect, perfectly, the requirements for a chemotherapeutic agent. The effects of vitamin C on cancer can be explained theoretically. Cell culture studies clearly demonstrate the killing of cancer cells. Animal studies show anticancer activity. In humans, measurements show that blood levels that would be expected to kill cancer cells can be reached, with intravenous injection. The blood plasma of people injected with vitamin C can kill cancer cells. There are a large number of clinical reports that ascorbate is an effective treatment. Case studies report people with metastatic cancer have been cured. The only item missing from this description is a randomised, double-blind, clinical trial showing intravenous ascorbate is an effective treatment or cure. The double-blind clinical trials by the medical establishment were performed by people who did not understand the difference between injected doses and tablets.

Of course, these studies do not *prove* that vitamin C is a *cure* for cancer, or even that it would provide a suitable treatment. However, such criticism is more than unscientific. The thoughtful reader will no doubt recognise a simple cost-benefit analysis. With powerful evidence of efficacy and a large degree of safety, there is little to lose and, potentially, a life to gain.

Replication and refutation

"Anyone who conducts an argument by appealing to Authority is not using his intelligence, he is just using his memory." Leonardo da Vinci

In previous chapters, we have described and explained the published data on the use of vitamin C as a therapeutic agent. We have also presented a dynamic flow model, to explain the apparently contradictory results obtained by different studies. The action of vitamin C in disease follows the law of mass action, proposed by Fred Klenner and extended by Robert Cathcart. The dynamic flow model is consistent with the claims of Linus Pauling and others, who have recommended megadose supplementation. It is also consistent with research that has showed little or no benefit from infrequently administered or inadequate doses of vitamin C.

According to the dynamic flow model, if vitamin C is to have a pharmacological effect, such as preventing infection or heart disease, blood levels must be raised consistently. As we have seen, a single two-gram dose will raise blood levels transiently, but average levels will not be much higher than with a single half-gram dose. Ascorbate is rapidly absorbed from the gut into the blood but is then excreted. Two separate one-gram doses, given several hours apart, will raise the average blood level for longer than a single, two-gram oral dose. Ideally, vitamin C supplementation should be based on multiple doses, taken at regular intervals throughout the day.

In most animals, large amounts of vitamin C are manufactured within the body. Alternatively, some animals obtain ascorbate from a diet of fruit and vegetables. In either case, its release into the blood stream is slower than if a tablet is taken. A daily megadose raises blood levels for less than half the day. If the aim is to prevent infection, this leaves long periods in which blood levels are too low. Since most experiments on vitamin C and disease have used single daily supplements, the variable results are only to be expected.

Oral megadose supplements of vitamin C are often inadequate as a treatment for disease. In some cases, Cathcart's method of titration to bowel tolerance may allow sufficient ascorbate to be taken. We have stressed the difference between oral vitamin C and intravenous sodium ascorbate. We can find no reason to discount the reported clinical observations, which suggest that intravenous ascorbate is effective as a treatment for heart disease, infection and cancer. Indeed, the reader may note that we have restricted our discussion of the multitude of diseases that vitamin C is claimed to prevent, treat or even cure. Even if the prospect of validation of each individual claim being confirmed were low, which it is not, the likelihood of some of them being confirmed is substantial.

In heart disease, cancer and infection, the claims for vitamin C are based on solid evidence. If the evidence were for an isolated branch of medicine or a single disease, it would be less compelling. However, the hypotheses are strong, easily testable and cover major killers in the developed world. If any one of these claims were ultimately established, the benefits would be considerable. Moreover, if even a few of the claims prove to be correct, the advance would amount to a medical revolution.

Hypotheses

A number of hypotheses and evidence-based theories concerning the action of vitamin C in health and disease require urgent investigation. It may seem a little strange that this book ends with a statement of hypotheses. However, from the scientific method presented earlier, it will be clear that this is inevitable. Science does not prove the case but enables the generation of hypotheses: these allow us to test the things we think we know, and pose the questions that need answers. The implications of current evidence make disturbing reading. It is therefore critical that the following hypotheses are tested as soon as possible.

Atherosclerosis is a form of scurvy

This hypothesis suggests that supplementation with high dose vitamin C will prevent heart disease and occlusive stroke in man. Experimental and clinical results suggest that shortage of ascorbate may be the primary cause of heart disease. Since this one of the biggest killers in the industrialised world, we might expect the medical profession to test this hypothesis impartially.

Vitamin C is an effective treatment for atherosclerosis and related disease

The evidence suggests that vitamin C can stabilise and prevent plaque growth in people with heart disease. Vitamin C may cause plaques to shrink, partially reversing the disease. Sustained high dose vitamin C may prevent heart attacks in people at high-risk. These findings require valid replication of the early experiments.

Lysine and vitamin C shrink atherosclerotic plaques

Linus Pauling suggested that the amino acids lysine and proline combined with vitamin C could shrink atherosclerotic plaques and "cure" heart disease. While "cure" may be too strong a suggestion, based on the available evidence there could be a synergistic effect of lysine and vitamin C, in shrinking and stabilising plaque. This treatment provides a valid hypothesis, with a provisional mechanism of action.

Antioxidants can cure heart disease

Both vitamin C and the tocotrienol form of vitamin E have been shown to reverse atherosclerotic plaques. It needs to be established whether these results are reproducible. It is possible that vitamin C and tocotrienols will reverse and even cure atherosclerosis. Again, these studies should be carried out soon, to avoid further unnecessary deaths.

Sustained high dose vitamin C prevents infectious disease

There have been studies on vitamin C and the common cold, but the doses used were generally low and would not have sustained adequate blood levels. Supplementation in the dynamic flow range, where a preventative response is predicted, should be subject to appropriate experimental testing.

Massive doses of vitamin C are an effective treatment for infection

There are a large number of clinical reports of massive doses of intravenous sodium ascorbate being an effective treatment for infectious diseases. There are no solid clinical trials in this area. Ethically, the clinical reports of substantial benefits cannot be ignored and should be followed up by clinical trials.

Vitamin C is an effective anti-cancer agent

There is substantial evidence suggesting that ascorbate will kill cancer cells preferentially. The clinical reports of vitamin C in cancer treatment are entirely consistent with this and imply a requirement for intravenous administration. The reports of successful treatment of cancer with vitamin C have a scientific basis. This hypothesis has a high likelihood of clinical confirmation.

Vitamin C could form the basis for new cancer treatments

Both lipoic acid and vitamin K reportedly increase the effectiveness of ascorbate in killing cancer cells. That is, lower doses of ascorbate can be used when either vitamin K or lipoic acid is also administered. This suggests the intriguing possibility that the actions of vitamin K and lipoic acid are synergistic with those of vitamin C. It may therefore be possible to increase the reported effectiveness of vitamin C as a cancer treatment. One again, appropriate studies should be carried out, as a matter of urgency.

Conclusions

Since vitamin C is so safe, it is nonsensical not to use it for treating life-threatening diseases, where there is supporting evidence. The complete picture may take years to develop but the benefits clearly outweigh the risks. This is particularly true for untreatable diseases or those where the conventional treatment has vicious side effects, such as chemotherapy. The argument that the benefits have not been "proven" is unscientific; such a statement could always be applied, no matter how strong the evidence. Indeed, the conventional idea that large-scale clinical trials are required is scientifically unsupportable. In making decisions about whether or not treatment with vitamin C is worth trying, we need to consider the cost-benefit analysis.

Costs are not only financial, but include health costs such as side effects. Normal clinical practice means a drug needs to be shown to be safe before use. In the case of vitamin C, demonstrating safety is almost irrelevant. Vitamin C is a GRAS substance, which means it is *generally regarded as safe*. According to the US Food and Drug Administration, GRAS substances are those whose use is recognized as safe, based on their extensive history of use in food before 1958 or on published scientific evidence. People need vitamin C to live. Millions of people have taken gram level supplements for years, apparently without harm. Whenever the medical establishment has tried to show a problem with the safety of vitamin C for normal individuals, the evidence has been unsound.

A person with atherosclerosis, who is in danger of a stroke or heart attack, can ask their physician about the Pauling or antioxidant network therapy. It is hard to understand how physicians treating patients with cardiovascular disease can justify not recommending ascorbate and tocotrienols. The evidence is not complete, but vitamin C, lysine and tocotrienols are safe. On a cost-benefit analysis, it is a "no-brainer", offering a possible cure with no downside.

A person with a minor infectious disease can opt to supplement with high-dose vitamin C. There are few effective conventional treatments for viral diseases. In the extreme cases of Ebola and other haemorrhagic fever outbreaks, the expectation is death without treatment. Despite the fact that there is no alternative treatment, the medical profession have not tried high-dose ascorbate, as proposed by Cathcart and others, on life threatening viral diseases. Since intravenous ascorbate is so safe, Cathcart asks why doctors treating Ebola and other severe viral infections have not tried it. We would also like to know the answer to this question.

We cannot imagine a sensible defence to the establishment position regarding ascorbate and cancer. We simply suggest that people with terminal cancer should ask their physician to explain why they are not receiving ascorbate therapy. In the absence of randomised, double blind, clinical trials showing no effect or a negative response, to deny a cheap, innocuous but possibly life-saving treatment is unethical. It certainly does not conform to any reliable medical decision-making, scientific justification or cost-benefit analysis based on the patient's needs.

At the time of writing, there is an international plan, called the Codex Alimentarius, to restrict the availability of vitamins and other nutrients, based on assumptions of safety and trade. The successful implementation of laws based on the Codex will mean that people who believe in prevention of disease with vitamins and associated nutrients will be unable to supplement themselves. Some of the proposed restrictions could mean, for example, that tocotrienols are unavailable for people with heart disease. The situation reminds us of the account of child deaths by Dr Archie Kalokerinos. Death rates of up to 50%, in Australian children, were a direct result of shortage of vitamin C and continued despite identification of the cause.[575]

Restriction of supplements will clearly mean that any disease resulting from chronic scurvy or other deficiency will continue, by law. The result will be regular profits for the pharmaceutical and related medical industries, which are involved in the treatment and management of unnecessary illness and death. Some doctors and alternative medical experts have described this as genocide. This book has given you the information needed to decide for yourself.

"We are trying to prove ourselves wrong as quickly as possible, because only in that way can we find progress." Richard P. Feynman

Glossary of terms

Acetyl-L-carnitine. A form of the amino acid carnitine, a food supplement.

AIDS. Acquired immune deficiency syndrome; a disease that is thought to be caused by the HIV retrovirus.

Alpha-tocopherol. A form of vitamin E.

Amino acid. A chemical building block that is connected in chains to form proteins.

Anencephaly. A defect in the development of the neural tube, which can lead to the loss of much of the brain, scalp and skull, and is linked to a deficit of folic acid in the diet.

Anaerobic. Growing, living or occurring in the absence of oxygen.

Ankylosing spondylitis. A painful rheumatic disease, in which inflammation of the joints of the spine can cause fusion of the vertebrae and rigidity.

Antibiotic. A chemical that will selectively kill bacteria or other microorganisms without damaging host tissues. Some specialised antibiotics are used in the treatment of cancer.

Antibody. A blood protein of the immune system that can combine with a specific invading bacteria, virus or cancer cell.

Antioxidant. A molecule that can donate an electron; any substance or process that prevents oxidation.

Apolipoprotein. A protein found in lipoproteins, which are used to carry cholesterol around the body in the blood.

Ascorbate. Another name for vitamin C.

Ascorbic acid. An alternative name for vitamin C.

Ascorbyl radical. Oxidised ascorbate, which has donated one electron and is an unreactive free radical.

Atrophic. Wasted or wasting away.

Aorta. The largest artery in the body, which carries oxygenated blood from the left side of the heart to the main arteries, and hence to the body tissues.

Benign. A benign tumour is a form of cancer that is non-invasive and slow growing.

Cancer. A group of diseases in which cells grow out of control and can spread to other parts of the body.

Carcinogen. Something that can cause a cancer.

Carcinogenic. Cancer causing.

Carcinoma. A cancer originating in an endothelial or epithelial tissue.

Catalase. An antioxidant enzyme, which was formed early in evolution and is used to break down hydrogen peroxide into oxygen and water.

Catalyst. An agent that participates in and speeds up a chemical reaction but itself remains unchanged. Enzymes are biological catalysts.

Chain breaking antioxidant. An antioxidant that will stop free radical chain reactions.

Chemotaxis. Movement of a cell along a chemical gradient.

Chronic inflammation. Long-term inflammation, such as that responsible for atherosclerosis and much degenerative disease.

Collagen. A fibrous protein molecule that provides strength to tissues and bones.

Collagenase. An enzyme that breaks down collagen.

Complement. A group of blood proteins that assist antibodies and white blood cells to combat bacteria and other foreign bodies.

Cytotoxic. Harmful to cell structure and function, may kill cells.

Dehydroascorbate. An oxidised form of vitamin C, that has lost two electrons.

Differentiation. The process, during development, of cell specialisation.

Doppler effect. Change in wavelength of sound (or other radiation) with speed, an example is the changing sound (pitch) of a racing car as it passes by.

DNA. Deoxyribonucleic acid, a long molecule consisting of small subunits that encode genetic information. Genes are made of DNA.

Electron. A subatomic particle with a negative electric charge.

Electron donor. A molecule that can provide (donate) electrons to other molecules.

Encapsulation. The process of wrapping a tumour, for example, in fibrous connective tissue.

Enzyme. A protein that increases the rate of a chemical reaction; a biological catalyst.

Epithelium. The outer layer of cells.

Epithelial cell. A surface cell, a cell from the epithelium.

Erythrocyte. A red blood cell.

Experiment. A test of an idea, hypothesis, or theory.

Extracellular. Outside the cell.

Evolutionary fitness. The ability of a living form to thrive and reproduce in a particular environment.

Fatty acid. A long lipid molecule, that has both a hydrophilic (water loving) end and a hydrophobic (water hating) end.

Fibrin. A blood protein involved in forming clots.

Fibrinogen. A blood protein, which is a precursor to fibrin.

Fibrinolysis. The destruction of fibrin.

Fibroblast. A cell that can form collagen and is important in wound healing; it is usually elongated and often found in connective tissues.

Free radical. A highly reactive molecule, an atom or a molecule with an unpaired electron.

Free radical scavenger. An antioxidant that reacts with and neutralises free radicals.

Free radical theory of aging. Theory that aging is caused by the action of free radicals over time, especially those that damage mitochondria.

Gene. A length, or lengths, of DNA that encode a protein; the basic unit of heredity.

Glutathione. A small water-soluble antioxidant, normally present in cells. A small peptide, which is often considered the most important water-soluble antioxidant within cells.

Glycation. A process in which a carbohydrate is attached to another molecule.

Glycosaminoglycan. A polysaccharide that forms the ground substance for collagen fibres in connective tissue.

Ground substance. A matrix of collagen fibres in a glycosaminoglycan gel; the base glue that holds the body together.

HDL. High density lipoprotein, aka "good" cholesterol.

HIV. The human immunodeficiency virus, thought to be the cause of AIDS.

Haemorrhage. A bleed.

Haemoglobin. A protein molecule containing iron, used for transporting oxygen; haemoglobin is found in red blood cells and gives blood its red colour.

Homeostasis. One of biology's core control systems for the maintenance of a steady state, rather like a thermostatic control mechanism.

Hyaluronic acid. A glycosaminoglycan that is part of the ground substance in connective tissue. Hyaluronic acid binds cells together and helps lubricate joints.

Hyaluronidase. An enzyme that breaks down hyaluronic acid, thereby decreasing its viscosity and increasing the permeability of connective tissue. It is sometimes used in medicine to increase absorption and diffusion of drugs administered by injection.

Hydrogen peroxide. An important biological free radical (H_2O_2), related to oxygen and water.

Hydroxyl radical. A highly reactive free radical (HO·) that can damage biological tissues rapidly.

Hypothesis. An idea that can be tested by experiment, a source of new information in science.

Iatrogenic. A disease caused by a physician or medical treatment.

Immunocompetence. The effectiveness of the immune system to keep the body free from disease.

Immunoglobulin. An antibody.

Immunosuppression. A reduction in the capacity of the immune system to defend the body against invasion.

Inflammation. A bodily response to injury, a general defensive reaction found in degenerative disease; inflamed tissues may be reddened, swollen or itchy.

Inflammatory cell. A cell involved in the process of inflammation, such as a macrophage or neutrophil.

Inflammatory messenger. A chemical released by inflammatory or inflamed cells, promoting an inflammatory response.

Ionisation. The process of forming an ionic (charged) molecule.

Ionising radiation. Radiation, such as x-rays and gamma-rays, that can remove electrons from atoms or molecules.

Intracellular. Inside the cell.

In vitro. Experimental – literally "in glass", as in a test tube.

In vivo. In the body.

Leukaemia. Strictly speaking, refers to cancer of the white blood cells (the leukocytes) but, in practice, it can be used to signify cancer of any cells in the blood or bone marrow, as in red cell leukaemia (erythroleukaemia).

Leukocyte. A white blood cell of any type.

LDL. Low-density lipoprotein, aka "bad" cholesterol.

Lipid. A fat or oil.

Lipoprotein. A substance made of lipid and protein.

Lipoic acid. A small, water- and lipid-soluble, sulphur containing antioxidant.

Lymph. A drainage system for tissues.

Lymph gland. A node or swelling in the lymphatic system.

Lymph node. A lymph gland.

Lymphocyte. A small white blood cell (leukocyte) formed in the lymphatic system; often present in inflammation.

Lymphadenopathy. Enlarged lymph glands.

Lysine. An essential amino acid in the diet, used to make protein; may be involved in heart disease.

Macromolecules. Very large biological molecules, such as proteins or DNA

Macrophage. A white blood cell that can engulf or eat invading bacteria.

Malignant. A tumour that is growing and invading tissues.

Megadose. A large dose of a vitamin or other food supplement, a gram of vitamin C or more; usually refers to large nutritional doses rather than pharmaceutical level doses.

Metabolism. The biochemical reactions that create energy from the oxidation of food.

Metastasis. Secondary cancers that have formed away from the site of the original tumour.

Mitochondria. Small structures, found in cells, that metabolise glucose and other substances to provide the cell with energy.

Mitochondrial leakage. Free radicals produced by the mitochondria during normal metabolism, and when the mitochondria are damaged.

Mitochondrial theory of aging. The idea that aging is caused by free radical damage to mitochondria over time.

Monocyte. A white blood cell with a single nucleus; a monocyte can engulf bacteria and foreign matter.

Myelogenous leukaemia. A condition in which the bone marrow makes too many white blood cells.

NAD. Nicotinamide adenine dinucleotide; a molecule involved in redox reactions during normal metabolism.

NADH. A reduced form of NAD.

NADP. Nicotinamide adenine dinucleotide phosphate; a redox molecule found in cells and related to NAD.

NADPH. A reduced form of NADP.

Necrosis. The process of tissue death; necrotic tissue is dead or dying.

Nucleotides. Small molecules that are linked together to form DNA or RNA.

Nucleic acid. A component of genes, such as DNA or RNA.

Nucleus. The central part of a cell, a structure in the cell that contains the DNA, bound to proteins.

Neutrophil. A white blood cell, that can engulf and destroy bacteria; involved in inflammation.

Organ. A structure in the body, for example: kidney, heart, liver.

Organelle. A subcellular structure, for example, a mitochondrion.

Orthomolecular. Orthomolecular medicine is the clinical use of substances that occur normally in the human body.

Osteoclast. A large multinucleate cell that absorbs calcified bone. Osteoclasts are found where bone is being absorbed and are often found in small depressions on the surface of bone.

Oxidation. The loss of an electron from a molecule; the opposite of reduction.

Oxidative damage. Damage to cells or tissues caused by oxidation by free radicals.

Oxidative stress. Stress or damage to cells and tissues caused by an increase in oxidising free radicals.

Oxygen. A gaseous molecule, O_2, which is essential to life but is also a deadly poison when pure or under pressure.

Oxygen poisoning. Damage caused by an excess of oxygen, which is toxic as it produces free radicals in tissues, in a process similar to radiation poisoning.

Periaortic. Around the aorta.

Phagocyte. A cell that can engulf bacteria or substances recognised as foreign.

Plaque. A pathological build up of protein and fats within the artery wall.

Plasma. The colourless fluid part of blood, lymph, or milk, in which corpuscles or fat-globules are suspended. It is also used as a term for extra-cellular and intra-cellular fluid.

Platelet. A microscopic disc-like structure in the blood, involved in blood clotting.

Polyunsaturated fat. A fat molecule with a relatively large number of double bonds between its carbon atoms.

Prostaglandin. A type of hormone.

Pro-oxidant. A substance that causes or promotes oxidation, the opposite of an antioxidant. Vitamin C acts mostly as an antioxidant, but can act as a pro-oxidant in the presence of iron.

Protein. A large molecule made up of a string of amino acids connected together; also known as a polypeptide.

Proton. A hydrogen atom that has lost its electron, a positively charged subatomic particle.

Reactive species. An active molecule that participates in free radical reactions; includes ROS (reactive oxygen species) and RNS (reactive nitrogen species).

Redox signalling. Biological communication using free radicals or reactive species for transmission of information.

Reducing agent. A molecule that can reduce other molecules by giving them one or more electrons, an antioxidant.

Reduction. The process of donating an electron, the opposite of oxidation.

Regression. As applied to cancers: shrinking rather than growing.

Remission. A reduction in the symptoms or activity of a disease; may be temporary.

Redox. The redox potential relates the **red**ucing and **ox**idising factors of the system.

Refutation. The process of showing that a scientific theory is incorrect.

Replication. The process of repeating an experiment.

Retrovirus. An RNA containing virus, which makes DNA using reverse transcriptase when it infects a cell.

Reverse transcriptase. An enzyme that makes DNA, using RNA as a template.

RNA. A nucleic acid similar to DNA but differing in one of its nucleotides. RNA is used instead of DNA in retroviruses, such as HIV.

Sarcoma. Cancer originating in the connective tissues.

Saturated fat. A fat molecule with no double bonds between its carbon atoms.

Semidehydroascorbate. The ascorbyl radical; oxidised ascorbate that has donated an electron to another molecule.

Spina bifida. A neural tube defect in which the spinal cord is exposed. This defect is linked to deficiency in folic acid.

Staphylococcus. A type (genus) of spherical bacteria, some of which are implicated in human diseases, such as boils and abscesses.

Streptococcus. A type (genus) of spherical anaerobic bacteria, some of which can live in the respiratory and digestive system. Scarlet fever is caused by Streptococcus pyogenes.

Stress protein. A protein produced by cells in response to stress.

Superoxide. A mildly reactive form of oxygen with an additional electron, O_2^-.

Thiol. A sulphur containing chemical group; used to describe molecules that contain such groups in their structure: lipoic acid is a thiol antioxidant. Thiols are important in signalling the antioxidant state of the cell.

Tissue. A group of similar cells, united to perform a specific function.

Toxin. A poison.

Tumour. A cancerous lump.

Tumour necrosis factor. A protein involved in inflammation, discovered as a substance that killed tumours.

Theory. An idea or hypothesis that has some theoretical or experimental support.

Tocopherol. Vitamin E consists of a set of molecules called tocopherols, that are fat soluble antioxidants. (See also tocotrienols.)

Tocotrienol. The tocotrienols are a form of vitamin E with antioxidant molecules that are more flexible and mobile than the tocopherols.

Ureter. A small tube leading from each kidney to the bladder.

Vitamin. An essential part of the diet, originally considered to be required in small amounts and known as a micronutrient. Some vitamins, like ascorbate, may be required in large amounts.

Index

AIDS
   remission with vitamin C
      174
animals
   reversing athersclerosis
      151
antibodies   188
ascorbate
   as a drug   124
   ratio to
      dehydroascorbate   118
atherosclerosis   133, 138
   as scurvy   212
   oxygen theory   142
   reversal in humans   151
bias
   and publicity   99
   control of   24
   experimental   23
   experimenter   25
   in medicine   16
   in presentation   24
bioavailability   85
biological variation   76
bowel tolerance
   and Cathcart   122
   of vitamin C   96
Brighthope Ian
   and AIDS   184
Cameron Ewan
   cancer studies   193
cancer   13
   and cytotoxic drugs   193
   benign   191
   carcinoma   192
   cured with vitamin C   197
   malignant   191
   Mayo Clinic study   203
   sarcoma   192
   treatment with vitamin C
      197
case studies   27
Cathcart Robert F.
   and AIDS   184
   theory of action   117
cell death
   apoptosis   110
   autoschizis   208
cholesterol   134
   and homeostasis   137
   good and bad   134
   hypercholesterolaemia
      (familial)   135
   hypothesis   134
   in diet   138
clinical experience
   infection   125
clinical trial
   definition   26
   double-blind   26
   large scale   33
collagen   20
common cold   176
   pharmacology and
      nutrition   179
   prevention   178
Cousins
   Norman   16
dehydroascorbate
   as oxidised ascorbate   65
   ratio to ascorbate   118
diabetes   109
   source of oxidation   118
disease

ascorbate -
   dehydroascorbate ratio
                                   121
   prevention   39
   treatment   39
dose
   and extrapolation   34
   level and repetition 179
   nutritional   105
dynamic flow model   91,
  127
   disease prevention   130
   intake levels   130
ebola
   suggested treatment 186
electrons
   and oxidation   58
evolution   69
experiment
   bad design   103
   definition   19
foam cells *See* white blood
   cells
folic acid   45
free radical
   as oxidising agent   59
   definition   59
   theory of aging   66
gene mutation
   and vitamin C   70
glucose
   see sugar   139
glutathione   115
glycation   139
guinea pig
   heart disease and
      vitamin C   149
HDL   See cholesterol
heart disease   12
   and free radicals   142
   and infection   141
   as atherosclerosis 133

   as scurvy   145
   reversal   164
Heart Protection Study 102
homocysteine   142
hydrogen peroxide 63
hydroxyl radical   62
hypoascorbemia   72
hypothesis
   definition   21
   null   22
Ignarro Louis J.
   vitamin C and arteries
                    149
infectious disease 12, 173
inflammation
   and antioxidants   115
   and free radicals   111
intravenous ascorbate
   and cancer   202
   risks   209
kidney stones
   and vitamin C   94
Klenner Fred
   and infection   175
   and polio   174
   law of mass action 118
   polio   182
Kuhn Thomas
   scientific revolutions 41
Langmuir Irving M.
   and suspect science 32
law
   definition   22
LDL   See cholesterol
least risk of inadequacy 75
least risk of toxicity   75
Levine Mark   77
Lind James
   experiment   20
lysine
   and the Pauling Therapy
                    158

medical establishment 14, 15
megadoses
  as nutritional amounts 117
metabolism
  limited supply of antioxidants 118
mitochondria
  and antioxidants 66
  and free radicals 67
Moertel Charles 15
  cancer experiments 200
Newton-Smith William 36
nitrogen
  reactive 61
numbers-needed-to-treat
  definition 35
Nutrition
  and pharmacology 105
observational method 27
oxidation 59
oxygen 57
  poisoning 61
  reactive 61
Oxygen
  and oxidation 58
Patterson J.C
  heart disease experiments 147
Pauling Linus
  and common cold 35
  as a scientist 54
  cancer studies 193
  cure for heart disease 157
  updated therapy 169
penicillin
  discovery of 27
placebo 25
plaque
  and arterial stress 148
  as inflammation 140
  development 139
  location in arteries 140
poliomyelitis See Klenner Fred
Popper Karl
  and refutable knowledge 50
profits
  and patents 42
proof
  in mathematics 19
  in science 19
RDA
  biochemical basis of 77
reduction 59
refutation
  definition 19
replication
  definition 19
  ease of 29
Riordan H.D and N.H.
  cancer and vitamin C 206
Sabin Albert B.
  polio vaccine 181
saturation
  misleading definition 82
science
  and fallibility 31
  suspect 32
scientific method 19
scientific revolutions 41
scurvy
  and James Lind 20
  prevention 37
Semmelweis
  and child bed fever 47
snake oil 45
spontaneous recovery 194
statins 102
Stone Irwin 54, 72
  cancer and vitamin C 193

stroke
    as atherosclerosis   133
sugar   139
superoxide   64
    production by white
        blood cells   113
Szent-Gyorgyi Albert
    discovery of vitamin C 52
theory
    aceptability and validity
        49
    characteristics of good 36
    definition   22
    good theory rules   36
tocotrienols
    as cure for heart disease
        163
ulcers
    of stomach   48
virus
    and vitamin C   175
    inactivation with vitamin C   174
vitamin B
    B1 (thiamine)   51
    B12 (cobalamin)   94
    B2 (riboflavin)   168
    B3 (niacin)   51
    B6 (pyridoxine)   142
    B7 (biotin)   47
    complex   47
vitamin C
    absorption and excretion   77
    actual requirements   91
    as cytotoxic drug   206
    as oxidant   66
    controversy   17
    deficiency   71
    dose   213
    history of   51
    oxidation in disease   118
    plasma levels   79
    safety   45, 93
    structure   64
    toxicity   39
    white blood cell levels 83
vitamin D   52
vitamin E
    as cure for heart disease   163
    tocopherols   104
    tocotrienols   163
    wrong form   104
vitamin K3
    and cancer   208
white blood cells
    lymphocytes   83
    macrophages   112
    monocytes   83
    neutrophils   83, 112
    phagocytes   112
Willis C.G.
    and heart disease   148

References

[1] Gigerenzer G. (2002) Calculated Risks, Simon and Shuster, New York.

[2] Pauling L. (1986) How to live longer and feel better, Avon Books, New York.

[3] Hoffer A. Vitamin C and Cancer, Quarry Health Books, Quebec, Canada.

[4] Cousins N. (1979) Anatomy of an Illness, Bantam Books, New York.

[5] Williams R.J. (1998) Biochemical Individuality: The Basis for the Genetotrophic Concept, Keats Publishing, Connecticut.

[6] Huff D. (1954) How to Lie with Statistics, W. W. Norton, New York.

[7] Shapiro A.K. (1977) The Powerful Placebo, Johns Hopkins Univ. Press, Baltimore.

[8] Rigway P.F. Darzi A.W. (2002) Placebos and standardising new surgical techniques, British Medical Journal, 325, 560.

[9] Fudenburg D. Tirole J. (1991) Game Theory, The M.I.T. Press, Massachusetts.

[10] Horrobin D.F. (2003) Are large clinical trials in rapidly lethal diseases usually unethical? Lancet, 361, 695-697.

[11] Pauling L. (1992) My love affair with vitamin C, Linus Pauling Papers, National Library of Medicine, USA.

[12] Barnett A. (2003) Revealed: how drug firms 'hoodwink' medical journals, Guardian Online, Sunday December 7.

[13] Newton-Smith W.H. (1981) The Rationality of Science, Routledge and Kegan, London

[14] Cathcart R.F. (1985) Vitamin C: The Nontoxic, Nonrate-limited, antioxidant free radical scavenger, Medical Hypotheses, 18, 61-77

[15] T. Kuhn (1962) The Structure of Scientific Revolutions, University of Chicago Press.

[16] Moynihan R. (2002) Alosteron: a case study in regulatory capture or a victory for patients rights? British Medical Journal, 325, 592-595.

[17] Lievre M. (2002) Alosteron for irritable bowel syndrome, 325, 555-556.

[18] Moynihan R. (2002) FDA advisors warn of more deaths if drug is relaunched, British meical journal, 325, 561.

[19] Keevil J. Lloyd C. Coulter J.L.S. (1957-63) Medicine and the Navy, 1200-1900, Livingstone, Edinburgh, 4 vols.

[20] Harvie D.I. (2002) Limeys, Sutton Publishing, England.

[21] Erasmus U. (1993) Fats That Heal, Fats That Kill, Alive Books, Canada.

[22] Kunin R.A. (1989) Snake oil, West J Med, 151(2), 208.

[23] Smithells R.W. Sheppard S. Schorah C.J. Seller M.J. Nevin N.C. Harris R. Read A.P. Fielding D.W. (1981) Apparent prevention of neural tube defects by periconceptional vitamin supplementation, Arch Dis Child, 56(12), 911-918.

[24] Smithells R.W. Nevin N.C. Seller M.J. Sheppard S. Harris R. Read A.P. Fielding D.W. Walker S. Schorah C.J. Wild J. (1983) Further experience of vitamin supplementation for prevention of neural tube defect recurrences, Lancet, 1(8332), 1027-1031.

[25] Mathews T.J. Margaret A. Honein M.A. David Erickson J.D. (2002) Spina Bifida and Anencephaly Prevalence, United States, 1991-2001, Morbidity and Mortality Weekly Report, September 13, 51(RR13), 9-11.

[26] Michie C.A. (1991) Neural tube defects in 18th century, The Lancet, 337, 504.

[27] Hibbard E.D Smithells R.W. (1965) Folic acid metabolism and human embryopathy, The Lancet, 12 June, 1254.

[28] Zempleni J. Mock D.M. (2000) Marginal biotin deficiency is teratogenic, Proc Soc Exp Biol Med, 223(1), 14-21.

[29] Loudon I. (2004) Ignaz Phillip Semmelweis (1818-1865). The James Lind Library (www.jameslindlibrary.org).

[30] Bishop W.J. The Early History of Surgery, Oldbourne Science Library, Olbourne Book Co, London, England.

[31] Foreman R. (1981) Medical resistance to innovation, Medical Hypotheses, 7, 8, 1981, 1009-1017.

[32] Popper K.R. (1983) A Pocket Popper, Ed by David Miller, Fontana, Oxford.

[33] Popper K.R. (1959) The Logic of Scientific Discovery, Hutchinson, London.

[34] Dunbar R. (1995) The Trouble With Science, Faber and Faber, London.

[35] Perutz M. (1998) I wish I'd made you angry earlier, Oxford University Press, Oxford, England.

[36] Halliwell B. and Gutteridge J.M.C. (1999) Free Radicals in Biology and Medicine, OUP, Oxford, England.

[37] Packer L. Colman C. (1999) The Antioxidant Miracle, Wiley, New York.

[38] Buettner G.R. Jurkiewicz B.A. (1993) The ascorbate free radical as a marker of oxidative stress: An EPR study, Free Radic Biol Med, 14, 49-55.

[39] Hossain M.A. Asada K. (1985) Monodehydroascorbate reductase from cucumber is a flavin adenine dinucleotide enzyme, J Biol Chem, 260, 12920-12926.

[40] Buttener G.R. (1993) The pecking order of free radicals and antioxidants: lipid peroxidation, alpha-tocopherol and ascorbate, Arch Biochem Biophys, 300, 535-543.

[41] Buettner G.R. Jurkiewicz B.A. (1996) Catalytic metals, ascorbate, and free radicals: combinations to avoid, Rad Research, 145, 532-541.

[42] Davies I. (1983) Aging, Institute of Biology Series, Arnold, London, England.

[43] Turrens J.F. (1997) Superoxide production by the mitochondrial respiratory chain, Biosci Rep, 17, 3.

[44] Liu J. Head E. Gharib A.M. Yuan W. Ingersoll R.T. Hagen T.M. Cotman C.W. Ames B.N. (2002) Memory loss in old rats is associated with brain mitochondrial decay and RNA/DNA oxidation: Partial reversal by feeding acetyl-L-carnitine and/or R-alpha - lipoic acid, Proc Natl Acad Sci, USA, 99(4), 2356-2361.

[45] Hagen TM, Liu J, Lykkesfeldt J, Wehr CM, Ingersoll RT, Vinarsky V, Bartholomew JC, Ames BN. (2002) Feeding acetyl-L-carnitine and lipoic acid to old rats significantly improves metabolic function while decreasing oxidative stress, Proc Natl Acad Sci, USA, 99(4), 1870-1875.

[46] Liu J. Killilea D.W. Ames B.N. (2002) Age-associated mitochondrial oxidative decay: Improvement of carnitine acetyltransferase substrate-binding affinity and activity in

[46] brain by feeding old rats acetyl-L- carnitine and/or R-alpha -lipoic acid, Proc Natl Acad Sci, USA, 19, 99(4), 1876-1881.

[47] Stone I. (1966) On the Genetic Etiology of Scurvy, Acta Geneticae Medicae et Gemellologiae, 15, 345-350.

[48] Challem J.J. Taylor E.W. (1998) Retroviruses, ascorbate, and mutations, in the evolution of Homo sapiens, Free Radic Biol Med, 25(1): 130-132.

[49] Bourne G. Allen R. (1935) Vitamin C in lower organisms, Nature, 136, 185-186.

[50] Stone I. (1972) The Natural History of Ascorbic Acid in the Evolution of the Mammals and Primates and Its Significance for Present Day Man, Orthomolecular Psychiatry, 1, 2 & 3, 82-89.

[51] Elwood S. McCluskey (1985) Which Vertebrates make vitamin C? Origins, 2(2), 96-100.

[52] Salomon L. L. Stubbs D. W. (1961) Some aspects of the metabolism of ascorbic acid in rats, Ann. N. Y. Acad. Sci., 92, 128-140.

[53] Conney A. H. et al. (1961) Metabolic interactions between ascorbic acid and drugs, Ann. N. Y Acad. Sci., 92, 115-127.

[54] Armour J. Tyml K. Lidington D. Wilson J.X. (2001) Ascorbate prevents microvascular dysfunction in the skeletal muscle of the septic rat, J Appl Physiol, 90(3), 795-803.

[55] Conney A. H. Bray C. A. Evans C. Burns J. J. (1961) Metabolic interactions between l-ascorbic acid and drugs, Ann N Y Acad. Sc, 92, 115-127.

[56] Pauling L. (1970) Evolution and the need for ascorbic acid.. Proc. Nat. Acad. Sc, USA, 67, 1643-1648.

[57] Pauling L. (1968) Orthomolecular Psychiatry, Science, 160, 265-271.

[58] Cathcart R.F (2003) Personal communication to Dr S. Hickey.

[59] Stone I. (1966) Hypoascorbemia, the genetic disease causing the human requirement for exogenous ascorbic acid, Pers. Biol. Med. 10, 133-134.

[60] Stone I. (1967) The Genetic Disease, Hypoascorbemia A Fresh Approach to an Ancient Disease and Some of its Medical Implications, Acta Geneticae Medicae et Gemellologiae, 16, 1, 52-60.

[61] Stone I. (1979) Eight decades of scurvy, the case history of a misleading dietary hypothesis, Orthomolecular Psychiatry, 8, 2, 58-62.

[62] Expert Group on Vitamins and Minerals (1999), UK government update paper EVM/99/21/P.

[63] Levine M. Conry-Cantilena C. Wang Y. Welch R. W. Phillip W. Washko P.W. Dhariwal K.R. Park J.B. Lazarrev A. Graumlich J.F. King J. Cantilena L.R. (1996) Vitamin C pharmacokinetics in healthy volunteers: Evidence for a recommended dietary allowance, Proc. Natl. Acad. Sci. USA, 93, 3704–3709.

[64] Standing Committee on Dietary reference Intakes (2000), Dietary Reference Intakes for Vitamin C, Vitamin E, Selenium, and Carotenoids: A Report of the Panel on Dietary Antioxidants and Related Compounds, Institute of Medicine, National Acadamy Press, USA.

[65] Levine M. Wang Y. Padayatty S.J. Morrow J. (2001) A new recommended dietary allowance of vitamin C for healthy young women Proc. Natl. Acad. Sci. USA, 14, 98 (17), 9842-9846.

[66] Kallner A. Hartmann I. Hornig D. (1979) Steady-state turnover and body pool of ascorbic acid in man. American Journal of Clinical Nutrition, 32, 530-539.

[67] Baker E.M. Hodges R.E. Hood J. Sauberlich H.E. March S.C. (1969) Metabolism of ascorbic-1-14C acid in experimental human scurvy. American Journal of Clinical Nutrition, 22, No. 5, 549-558.

[68] Kallner A. Hartmann I. Hornig D. (1977) On the absorption of ascorbic acid in man. International Journal of Vitamin and Nutrition Research, 47, 383-388.

[69] Hornig, D.H. and Moser, U. (1981) The safety of high vitamin C intakes in man. In: Vitamin C (ascorbic acid). Eds. Counsell, J.N. and Hornig, D.H., Applied Science Publishers, New Jersey, 225-248.

[70] Kallner, A. Hartmann I. Hornig D. (1981) On the requirements of ascorbic acid in man: steady-state turnover and body pool in smokers. American Journal of Clinical Nutrition, 34, 1347-1355.

[71] Young V.R. (1996) Evidence for a recommended dietary allowance for vitamin C from pharmacokinetics: A comment and analysis, Proc. Natl. Acad. Sci, USA, 93, 14344–14348.

[72] Ginter E. (2002) Current views of the optimum dose of vitamin C, Slovakofarma Revue XII, 1, 4-8.

[73] Padayatty S.J. Sun H. Wang Y. Riordan H.D. Hewitt S.M. Katz A. Wesley R.A. Levine M. (2004) Vitamin C pharmacokinetics: implications for oral and intravenous use, Ann Intern Med, 140(7), 533-537.

[74] Benke K.K. (1999) Modelling Ascorbic Acid Level in Plasma and Its Dependence on Absorbed Dose, Journal of the Australasian College of Nutritional & Environmental Medicine, 18(1), 11-12.

[75] Hornig D. (1975) Distribution of ascorbic acid, metabolites and analogues in man and animals, Ann NY Acad Sci, 258, 103-118.

[76] Moser U. (1987) The uptake of ascorbic acid by leukocytes, Ann NY Acad Sci, 198, 200-215.

[77] Watson R.W.G (2002) Forum Review, Redox Regulation of Neutrophil Apoptosis, Antioxidants and redox signaling, 4, 1, 97-104.

[78] Kinnula V.L. Soini Y. Kvist-Makela K. Savolainen E. Koisinen P. (2002) Antioxidant Defense Mechanisms in Human Neutrophils, Antioxidants and redox signaling, 4, 1, 27-34.

[79] Daruwala R. Song J. Koh W.S. Rumsey S.C. Levine M. (1999) Cloning and functional characterization of the human sodium-dependent vitamin C transporters hSVCT1 and hSVCT2, FEBS Lett, 5, 460(3), 480-484.

[80] Tsukaguchi H. Tokui T. Mackenzie B. Berger U.V. Chen X.Z. Wang Y. Brubaker R.F. Hediger M.A. (1999) A family of mammalian Na+-dependent L-ascorbic acid transporters, Nature, 399, 70-75.

[81] Omaye S.T. Schous E.E. Kutnink M.A. Hawkes W.C. (1987) Measurement of vitamin C in blood components by high performance liquid chromatography, implications in assessing vitamin C status, Ann NY Acad Sci, 498, 389-401.

[82] Perret J.L. Lagauche D. Favier J.C. Rey P. Bigois L. Adam F. (2004) Scurvy in intensive care despite vitamin supplementation, Presse Med, 33(3), 170-171.

[83] Ruiz E. Siow R.C.M. Bartlett S.R. Jenner A.M. Sato H. Bannai S. Mann G.E. (2003) Vitamin C inhibits diethylmaleate-induced L-cystine transport in human vascular smooth muscle cells, Free Radical Biology and Medicine, 34 (1) 103-110.

[84] Fletcher AE, Breeze E, Shetty P.S. (2003) Antioxidant vitamins and mortality in older persons: findings from the nutrition add-on study to the Medical Research Council Trial of Assessment and Management of Older People in the Community, Am J Clin Nutr, 78(5), 999-1010.

[85] Halliwell B. Zhao K. Whiteman M. (2001) The Gastrointestinal Tract: A Major Site of Antioxidant Action? Free Radic Res, 33(6), 819-830.

[86] Ekstrom A.M. Serafini M. Nyren O. Hansson L.E. Ye W. Wolk A. (2000) Dietary antioxidant intake and the risk of cardia cancer and noncardia cancer of the intestinal and diffuse types: a population-based case-control study in Sweden, Int J Cancer, 87(1), 133-140.

[87] Martin M. Ferrier B. Roch-Ramel F. (1983) Renal excretion of ascorbic acid in the rat: a micropuncture study, Am J Physiol. 1983 Mar; 244(3),335-341.

[88] Karlinsky N. (2002) Good Morning America, Sept 18.

[89] Baranowitz SA, Maderson PF (1995) Acetaminophen toxicity is substantially reduced by beta-carotene in mice, Int J Vitam Nutr Res, 65(3), 175-180.

[90] Hong R.W. Rounds J.D. Helton W.S. Robinson M.K. Wilmore D.W. (1992) Glutamine preserves liver glutathione after lethal hepatic injury, Ann Surg, 215(2), 114-119.

[91] Schnell R.C. Park K.S. Davies M.H. Merrick B.A. Weir S.W.(1988) Protective effects of selenium on acetaminophen-induced hepatotoxicity in the rat, Toxicol Appl Pharmacol, 95(1), 1-11.

[92] Ostapowicz G. Fontana R.J. Schiødt F.V. Larson A. Davern T.J. Han S.H.B. McCashland T.M. Shakil A.O. Hay J.E. Hynan L. Crippin J.S. Blei A.T. Samuel G. Reisch J. Lee W.M. (2002) Results of a Prospective Study of Acute Liver Failure at 17 Tertiary Care Centers in the United States, Annals of Internal Medicine, 137, 947-954.

[93] Gurkirpal S. (1998) Recent Considerations in Nonsteroidal Anti-Inflammatory Drug Gastropathy, The American Journal of Medicine, July 27, 31S.

[94] Wolfe M. Lichtenstein D. Gurkirpal S. (1999) Gastrointestinal Toxicity of Nonsteroidal Anti-inflammatory Drugs, N Engl J Med, 340(24) 1888-1889.

[95] Kohn L. Corrigan J. Donaldson M. (1999) To Err Is Human: Building a Safer Health System. Washington, DC: National Academy Press.

[96] Leape L.L.(1992) Unnecessary surgery, Annu Rev Public Health, 13, 363-383.

[97] Phillips D.P. Christenfeld N. Glynn L.M. (1998) Increase in US medication-error deaths between 1983 and 1993, Lancet 28, 351(9103), 643-644.

[98] Lazarou J. Pomeranz B.H. Corey P.N. (1998) Incidence of adverse drug reactions in hospitalized patients: a meta-analysis of prospective studies, JAMA, Apr, 15, 279(15), 1200-1205.

[99] Levine M, Rumsey SC, Daruwala R, Park JB, Wang Y (1999) Criteria and recommendations for vitamin C intake, JAMA, 281, 1415–1423.

[100] Stone I. Hoffer A. (1976) The Genesis of Medical Myths, Orthomolecular Psychiatry, 5, 3, 163-168.

[101] McCormick W.J. (1946) Lithogenesis and hypovitaminosis, Medical Record, 159, 410-413.

[102] Chalmers A.H, Cowley DM, Brown J.M. (1986) A possible etiological role for ascorbate in calculi formation, Clin Chem, 32(2), 333-336.

[103] Curhan, G. C., Willett, W. C., Speizer, F. E., Stampfer, M. J. (1999) Megadose Vitamin C consumption does not cause kidney stones. Intake of vitamins B6 and C and the risk of kidney stones in women, J Am Soc Nephrol., Apr, 10, 4, 840-845.

[104] Curhan G.C. Willett W.C. Rimm E.B. Stampfer M.J. (1996) A prospective study of the intake of vitamins C and B6, and the risk of kidney stones in men, J Urol, 155(6), 1847-1851.

[105] Johnston C.S. (1999) Biomarkers for establishing a tolerable upper intake level for vitamin C, Nutr Rev, 57, 71-77.

[106] Garewal H.S. Diplock A.T. (1995) How 'safe' are antioxidant vitamins? Drug Saf, Jul, 13(1), 8-14.

[107] Diplock A.T. (1995) Safety of antioxidant vitamins and beta-carotene, Am J Clin Nutr, 62(6 Suppl), 1510S-1516S.

[108] Dwyer J.H. Nicholson L.M. Shirecore A. Sun P. Noel C. Merz B. Dwyer K.M. (2000) Vitamin C supplement intake and progression of carotid atherosclerosis, the Los Angeles Atherosclerosis Study, American Heart Association 40th Annual Conference on Cardiovascular Disease Epidemiology and Prevention, La Jolla, March 2-3.

[109] Kritchevsky S.B. Shimakawa T. Tell G.S. Dennis B. Carpenter M. Eckfeldt J.H. Peacher-Ryan H. Heiss G. (1995) Dietary antioxidants and carotid artery wall thickness. The ARIC Study. Atherosclerosis Risk in Communities Study. Circulation, 15, 92(8), 2142-2150.

[110] Drossos GE, Toumpoulis IK, Katritsis DG, Ioannidis JP, Kontogiorgi P, Svarna E, Anagnostopoulos CE (2003) Is vitamin C superior to diltiazem for radial artery vasodilation in patients awaiting coronary artery bypass grafting? J Thorac Cardiovasc Surg, 125(2), 330-335.

[111] Anderson J.D. (1995) Computational Fluid Dynamics, McGraw-Hill, New York.

[112] Dwyer J. (2002,2004) Personal communication (email) to Dr Steve Hickey.

[113] Dwyer J.H. Paul-Labrador M.J. Fan J. Shircore A.M. Merz C.N. Dwyer K.M. (2004) Progression of carotid intima-media thickness and plasma antioxidants: the Los Angeles Atherosclerosis Study, Arterioscler Thromb Vasc Biol, 24(2), 313-319.

[114] Waters D.D. Alderman E.L. Hsia J. Howard B.V. Cobb F.R. Rogers W.J. Ouyang P. Thompson P.Tardif J.C. Higginson L. Bittner V. Steffes M. Gordon D.J. Proschan M. Younes N. Verter J.I (2002) Effects of Hormone Replacement Therapy and Antioxidant

Vitamin Supplements on Coronary Atherosclerosis in Postmenopausal Women, A Randomized Controlled Trial, JAMA, 288, 2432-2440.

[115] Heart Protection Study Collaborative Group. (2002) MRC/BHF Heart Protection Study of cholesterol lowering with simvastatin in 20,536 high-risk individuals: a randomised placebo-controlled trial, *Lancet*, Vol 360, 9326.

[116] Newman T.B, Hulley S.B. (1996) Carcinogenicity of lipid-lowering drugs, JAMA, 3, 275(1), 55-60.

[117] Ghirlanda G. Oradei A. Manto A. Lippa S. Uccioli L. Caputo S. Greco A.V. Littarru G.P. (1993) Evidence of plasma CoQ10-lowering effect by HMG-CoA reductase inhibitors: a double-blind, placebo-controlled study, J Clin Pharmacol, 33(3), 226-229.

[118] Watts G.F. Castelluccio C. Rice-Evans C. Taub N.A. Baum H. Quinn P.J. (1993) Plasma coenzyme Q (ubiquinone) concentrations in patients treated with simvastatin, J Clin Pathol, 46(11), 1055-1057.

[119] Tolbert J.A. (1990) Coenzyme Q.sub.10 with HMG-CoA reductase inhibitors, United States Patent 4,929,437, Assigned Merk & Co. (Rathway, NJ).

[120] Brown M.S. (1990) Coenzyme Q.sub.10 with HMG-CoA reductase inhibitors, United States Patent 4,933,165, Assigned Merck & Co. (Rahway, NJ).

[121] Arduini A. Peschechera A. Carminati P. (2001) Method of preventing or treating statin-induced toxic effects using L-carnitine or an alkanoyl L-carnitine, United States Patent 6,245,800, assigned Sigma-Tau (Rome, It).

[122] Whitaker J.M. (2002) Citizen petition to change the labelling for all statin drugs (mevacor, lescol, pravachol, zocor, lipitor, and advicor) recommending use of 100-200mg per day of supplemental coenzyme Q10 (including cardiomyopathy and congestive heart failure) to reduce the risk of statin induced myopathies, FDA petition, 24 May.

[123] Saldeen T. Li D. Mehta J.L. (1999) Differential effects of alpha- and gamma-tocopherol on low-density lipoprotein oxidation, superoxide activity, platelet aggregation and arterial thrombogenesis, J Am Coll Cardiol, 34(4), 1208-1215.

[124] Kontush A. Spranger T. Reich A. Baum K. Beisiegel U. (1999) Lipophilic antioxidants in blood plasma as markers of atherosclerosis: the role of alpha-carotene and gamma-tocopherol, Atherosclerosis, 144(1), 117-122.

[125] Ziouzenkova O. Winklhofer-Roob B.M. Puhl H. Roob J.M. Esterbauer H. (1996) Lack of correlation between the alpha-tocopherol content of plasma and LDL, but high correlations for gamma-tocopherol and carotenoids, J Lipid Res, 37(9), 1936-1946.

[126] Morton L.W. Ward N.C. Croft K.D. Puddey I.B. (2002) Evidence for the nitration of gamma-tocopherol in vivo: 5-nitro-gamma-tocopherol is elevated in the plasma of subjects with coronary heart disease, Biochem J, 15, 364(Pt 3), 625-628.

[127] Jiang Q. Elson-Schwab I. Courtemanche C. Ames B.N. (2000) gamma-tocopherol and its major metabolite, in contrast to alpha-tocopherol, inhibit cyclooxygenase activity in macrophages and epithelial cells, Proc Natl Acad Sci U S A, 97(21), 11494-11499.

[128] Ohrvall M. Sundlof G. Vessby B.J. (1996) Gamma, but not alpha, tocopherol levels in serum are reduced in coronary heart disease patients, Intern Med, 239(2), 111-117.

[129] Handelman G.J. Packer L. Cross C.E. (1996) Destruction of tocopherols, carotenoids, and retinol in human plasma by cigarette smoke, Am J Clin Nutr, 63(4), 559-565.

[130] Dietrich M. Block G. Norkus E.P. Hudes M. Traber M.G. Cross C.E. Packer L. (2003) Smoking and exposure to environmental tobacco smoke decrease some plasma antioxidants and increase gamma-tocopherol in vivo after adjustment for dietary antioxidant intakes, Am J Clin Nutr, 77(1), 160-166.

[131] Li D. Saldeen T. Romeo F. Mehta J.L. (1999) Relative Effects of alpha- and gamma-Tocopherol on Low-Density Lipoprotein Oxidation and Superoxide Dismutase and Nitric Oxide Synthase Activity and Protein Expression in Rats, J Cardiovasc Pharmacol Ther, 4(4), 219-226.

[132] Leonard S.W. Terasawa Y. Farese R.V.Jr, Traber M.G. (2002) Incorporation of deuterated RRR- or all-rac-alpha-tocopherol in plasma and tissues of alpha-tocopherol transfer protein-null mice, Am J Clin Nutr, 75(3), 555-560.

[133] Traber M.G. (1999) Utilization of vitamin E, Biofactors, 10(2-3), 115-120.

[134] Chopra R.K. Bhagavan H.N. (1999) Relative bioavailabilities of natural and synthetic vitamin E formulations containing mixed tocopherols in human subjects, Int J Vitam Nutr Res, 69(2), 92-95.

[135] Lauridsen C. Engel H. Craig A.M. Traber M.G. (2002) Relative bioactivity of dietary RRR- and all-rac-alpha-tocopheryl acetates in swine assessed with deuterium-labeled vitamin E, J Anim Sci, 80(3), 702-707.

[136] Muntwyler J. Hennekens C.H. Manson J.E. Buring J.E. Gaziano J.M. (2002) Vitamin supplement use in a low-risk population of US male physicians and subsequent cardiovascular mortality, Arch Intern Med, 162(13), 1472-1476.

[137] Simon J.A. (2002) Combined Vitamin E and Vitamin C Supplement Use and Risk of Cardiovascular Disease Mortality, Archives Internal Medicine, 162(22), Editors Correspondence.

[138] Gaziano M. Muntwyler J. (2002) Combined Vitamin E and Vitamin C Supplement Use and Risk of Cardiovascular Disease Mortality, Archives Internal Medicine, 162(22), In reply, Editors Correspondence.

[139] Cuzzocrea S., Riley D.P. Caputi A.P. Salvemini D. (2001) Antioxidant Therapy: A New Pharmacological Approach in Shock, Inflammation, and Ischemia/Reperfusion Injury, Pharmacological Reviews, 53(1), 135–159.

[140] Florence T.M. (1995) The role of free radicals in disease, Aust N Z J Ophthalmol, 23(1), 3-7.

[141] Bliss M. (1984) The Discovery of Insulin, Univ of Chicago Press, USA.

[142] Price K.D. Price C.S. Reynolds R.D. (2001) Hyperglycemia-induced ascorbic acid deficiency promotes endothelial dysfunction and the development of atherosclerosis, Atherosclerosis, 158(1), 1-12.

[143] Hunt J.V. (1995) Ascorbic acid and diabetes mellitus, In Subcellular Biochemistry, 25, Ascorbic acid: Biochemistry and Biomedical Cell Biology, ed Harris R.J., Plenum, New York.

[144] Buttke T.M. Sandstrom P.A. (1995) Redox regulation of programmed cell death in lymphocytes, Free Rad Res, 22, 389.

[145] Van Den Dobblesteen D.J. Nobel C.S. Schlegel J. Cotgreave I.A. Orrenius S. Slater A.F. (1996) Rapid and specific efflux of GSH during apoptosis induced by anti-FAS/APO-1 antibody, J Biol Chem, 271, 15420-15427.

[146] Farrow B. Evers B.M. (2002) Inflammation and the development of pancreatic cancer, Surg Oncol, 10(4), 153-169.

[147] Jaiswal M. LaRusso N.F. Gores G.J. (2001) Nitric oxide in gastrointestinal epithelial cell carcinogenesis: linking inflammation to oncogenesis, Am J Physiol Gastrointest Liver Physiol, 281(3), G626-634.

[148] Hsieh C.L. Yen G.C. (2000) Antioxidant actions of du-zhong (Eucommia ulmoides Oliv.) toward oxidative damage in biomolecules, Life Sci, 66(15), 1387-1400.

[149] Pricop L. Salmon J.E. (2002) Redox Regulation of Fcg Receptor-Mediated Phagocytosis, Implications for Host Defense and Tissue Injury, Antioxidants and Redox Signaling, 4, 1, 85-95.

[150] Swain S.D. Rohn Y .T. Quinn M.T. (2002) Neutrophil Priming in Host Defense: Role of Oxidants as Priming Agents, Antioxidants and Redox Signaling, 4, 69–83.

[151] Levine M. Dhariwal K.R. Wang Y. Park J.B. Welch R.W. (1994) Ascorbic Acid in Neutrophils, in Natural Antioxidants in Health and Disease, Ed Frei B., Academic Press, San Diego.

[152] Evans R.M. Currie L. Campbell A. (1982) The distribution of ascorbic acid between various cellular components of blood in normal individuals and its relation to the plasma concentration, B. J. Nutr, 17, 173-182.

[153] Jacob R.A. Pianalto F.S. Agee R.E. (1992) Cellular ascorbate depletion in healthy men, J. Nutr, 122, 1111-1118.

[154] Jariwallah R.J. Harekeh S. (1996) Antiviral and immunomodulatory activities of ascorbic acid, subcell Biochem, 25, 213-231.

[155] Wang Y. Russo T.A. Kwan O. Chanock S. Rumsey S.C. Levine M. (1997) Ascorbate recycling in human neutrophils: induction by bacteria, Proc Natl Acad Sci USA, 94, 13816-13819.

[156] Anderson R. Lukey P.T. (1987) A biological role for ascorbate in selective neutralisation of extracellular phagocyte derived oxidants, Ann NY Acad Sci, 498, 229-247.

[157] Halliwell B. Wasil M. Grootveld M. (1987) Biologically significant scavenging of the myeloperoxidase derived oxidant hypochlorous acid by ascorbic acid, Febs Lett, 213, 15-17.

[158] Heinecke J.W. (1997) Pathways for oxidation of low density lipoprotein by mycloperoxidase: tyrosyl radical, reactive aldehydes, hypochlorous acid and molecular chlorine, Biofactors, 6, 145-155.

[159] Lunec J. Blake D.R. (1985) the determination of dehydroascorbic acid and ascorbic acid in the serum and synovial fluid of patients with rheumatoid arthritis, Free Rad Reseacrh Commun, 1, 31-39.

[160] Cross C.E. Forte T. Stocker R. Louie S. Yamamoto Y. Ames B.N. Frei B. (1990) Oxidative stress and abnormal cholesterol metabolism in patients with adult respiratory distress syndrome, J Lab Clin Med, 115, 396-404.

[161] Lykkesfeldt J. Loft S. Neilsen J.B. Poulson I.I.E. (1997) Ascorbic acid and dehdroascorbic acid as biomarkers of oxidative stress caused by smoking, Am J Clin Nutr, 65, 959-963.

[162] Mudway I.S. Krishna M.T. Frew A.J. MacCleod D. Sandstrom T. Holgate S.T. Kelly F.J. (1999) Compromised concentrations of ascorbate in fluid lining the respiratory tract in human subjects after exposure to ozone, Occup Environ Med, 56, 173-181.

[163] Hazell L.J. Arnold L. Flowers D. Waeg G. Malle E. Stcoker R. (1996) presence of hypochlorite modified proteins in human atherosclerotic leasions, J Clin Invest, 97, 1535-1544.

[164] Byun J. Mueller D.M. Fabjan J.S. Heinecke J.W. (1999) Nitrogen dioxide radical generated by the myeloperoxidase hydrogen peroxide nitrite system promotes lipid peroxidation of low density lipoprotein, Febs Lett, 455, 243-246.

[165] Babior B.M. (1997) Superoxide: a two edged sword, Brazil J. Med Biol Res, 30, 141.

[166] Henderson L.M. Chappell J.B. (1996) NADPH oxidase of neutrophils, Biochim Biophys Acta, 87, 1273.

[167] 853 Ahluwalia J. Tinker A. Clapp L.H. Duchen M.R. Abramov A.Y. Pope S. Nobles M. Segal A.W. (2004) The large-conductance Ca2+-activated K+ channel is essential for innate immunity, Nature, 427, 853-857.

[168] Kettel A.J. Winterbourne C.C. (1997) Myeloperoxidase: a key regulator of neutrophil oxidant production, Redox Res, 3, 3.

[169] Weiss S.J. (1989) Tissue destruction by neutrophils, New Eng J Med, 320, 365.

[170] McCord J.M. Turrens J.F. (1994) Mitochondrial injury by ischemia and reperfusion, Curr Top Bioenerg, 17, 173.

[171] Horrobin D.F. (1996) Ascorbic acid and prostaglandin synthesis, Subcell Biochem, 25, 109-115.

[172] Chandrasekharan N.V. Dai H. Roos K.L. Evanson N.K. Tomsik J. Elton T.S. Simmons D.L. (2002) COX-3, a cyclooxygenase-1 variant inhibited by acetaminophen and other analgesic/antipyretic drugs: cloning, structure, and expression, Proc Natl Acad Sci U S A, 99(21), 13926-13931.

[173] Buettner G.R. (1993) The pecking order of free radicals and antioxidants, Lipid peroxidation,
a-tocopherol, and ascorbate, Arch Biochem Biophy, 300, 535-543.

[174] Holland J. (1998) Emergence: From Chaos to Order, OUP, Oxford, England.

[175] Klenner F. (1971) Observations On the Dose and Administration of Ascorbic Acid When Employed Beyond the Range of A Vitamin In Human Pathology, The Journal of Applied Nutrition, 23, 3 & 4, 61-88.

[176] Bors W. Buettner G.R. (1997) The vitamin C radical and its reactions, in Vitamin C in Health and Disease, Ed. by L. Packer and J. Fuchs, Marcel Dekker, New York, 75-94.

[177] Kubin A. Kaudela K. Jindra R. Alth G. Grunberger W. Wierrani F. Ebermann R. (2003) Dehydroascorbic Acid in urine as a possible indicator of surgical stress, Ann Nutr Metab, 47(1), 1-5.

[178] Sinclair A.J. Taylor P.B. Lunec J. Girling A.J. Barnett A.H. (1994) Low plasma ascorbate levels in patients with type 2 diabetes mellitus consuming adequate dietary vitamin C, Diabet Med, 11(9), 893-898.

[179] Rusakow L.S. Han J. Hayward M.A. Griffith O.W. (1995) Pulmonary oxygen toxicity in mice is characterized by alterations in ascorbate redox status, J Appl Physiol, 79(5), 1769-1776.

[180] Obrosova I.G. Fathallah L. Liu E. Nourooz-Zadeh J. (2003) Early oxidative stress in the diabetic kidney: effect of DL-alpha-lipoic acid, Free Radic Biol Med, 34(2), 186-195.

[181] Jiang Q. Lykkesfeldt J. Shigenaga M.K. Shigeno E.T. Christen S. Ames B.N. (2002) gamma-tocopherol supplementation inhibits protein nitration and ascorbate oxidation in rats with inflammation, Free Radic Biol Med, 33(11), 1534-1542.

[182] Simoes S.I. Eleuterio C.V. Cruz M.E. Corvo M.L. Martins M.B. (2003) Biochemical changes in arthritic rats: dehydroascorbic and ascorbic acid levels, Eur J Pharm Sci, 18(2), 185-189.

[183] Balestrasse K.B. Gardey L. Gallego S.M. Tomaro M.L. (2001) Response of antioxidant defence system in soybean nodules and roots subjected to cadmium stress, Australian Journal of Plant Physiology, 28(6), 497–504.

[184] Kuniak E. Skodowska M. (2004) The effect of *Botrytis cinerea* infection on the antioxidant profile of mitochondria from tomato leaves, Journal of Experimental Botany, 55(397), 605-612.

[185] Iheanacho E.N. Stocker R. Hunt N.H. (1993) Redox metabolism of vitamin C in blood of normal and malaria-infected mice, Biochim Biophys Acta, 1182(1), 15-21.

[186] Bhaduri J.N. Banerjee S. (1960) Ascorbic acid, dehydroascorbic acid and glutathione levels in blood of patients suffering from infectious diseases, Indian Journal of Medical Research, 48, 208-211.

[187] Chakrabarti B. Banerjee S. (1955) Dehydroascorbic acid level in blood of patients suffering from various infectious diseases, Proc Society Experimental Biology and Medicine, 88, 581-583.

[188] Du W.D. Yuan Z.R. Sun J. Tang J.X. Cheng A.Q. Shen D.M. Huang C.J. Song X.H. Yu X.F. Zheng S.B. (2003) Therapeutic efficacy of high-dose vitamin C on acute pancreatitis and its potential mechanisms, World J Gastroenterol, 9(11), 2565-2569.

[189] Jacob R.A. (1995) the integrated antioxidant system, Nutrition Research, 15, 755-766.

[190] Lewin S. (1976) Vitamin C: Its Molecular Biology and Medical Potential, Academic Press.

[191] Cathcart R.F. (1991) A unique function for ascorbate, Medical Hypotheses, 35, 32-37.

[192] Pauling L. (1988) General Chemistry, Dover, New York.

[193] May J.M. Qu Z. Li X. (2001) Requirement for GSH in recycling of ascorbic acid in endothelial cells, Biochemical Pharmacology, 62(7), 873-881.

[194] Vethanayagam J.G. Green E.H. Rose R.C. Bode A.M. (1999) Glutathione-dependent ascorbate
recycling activity of rat serum albumin, Free Radical Biology & Medicine, 26, 1591-1598.

[195] Mendiratta S. Qu Z.C. May J.M. (1998) Enzyme-dependent ascorbate recycling in human erythrocytes: role of thioredoxin reductase, Free Radical Biology & Medicine, 25, 221-228.

[196] Cathcart R.F. (1975) Clinical trial of vitamin C, Letter to the Editor, Medical Tribune, June 25.
[197] Cathcart R.F. (1981) Vitamin C titrating to bowel tolerance, anascorbemia and acute induced scurvy, Medical Hypotheses, 7, 1359-1376.
[198] Cathcart R.F. (1992) Unpublished paper on AIDS, letter to editor, Critical Path Project.
[199] Daruwala R. Song J. Koh W.S. Rumsey S.C. Levine M. (1999) Cloning and functional characterization of the human sodium-dependent vitamin C transporters hSVCT1 and hSVCT2, FEBS Letters, 460, 480-484.
[200] Ravskow U. (2002) Is atherosclerosis caused by high cholesterol? Q J Med, 95, 397-403.
[201] Ravnskov U. (2001) The Cholesterol Myths: Exposing the Fallacy That Cholesterol and Saturated Fat Cause Heart Disease, New Trends, Winona Lake, In, USA.
[202] Stehbens W.E. (2001) Coronary heart disease, hypercholesterolaemia, and atherosclerosis, Exp Mol Pathol, 70(2), 103-119.
[203] Stehbens W.E. Martin M. (1991) The vascular pathology of familial hypercholesterolaemia, Pathology, 23(1), 54-61.
[204] Sijbrands E.J.G. Westendorp R.G.J. Defesche J.C. de Meier P. H.E.M. Smelt A.H.M. Kastelein J.J.P. (2001) Mortality over two centuries in large pedigree with familial hypercholesterolaemia: family tree mortality study, BMJ, 322, 1019-1023.
[205] Ginter E. Bobek P. Kubec F. Vozar J. Urbanova D. (1982) Vitamin C in the control of hypercholesterolemia in man, Int J Vitam Nutr Res Suppl, 23, 137-152.
[206] Spittle C.R. (1971) Atherosclerosis and vitamin C, Lancet, 1280-1281.
[207] Morin R.J. (1972) Arterial cholesterol and vitamin C. Lancet, 594-595.
[208] Crawford G.P. Warlow C.P. Bennett B. Dawson A.A. Douglas A.S. Kerridge D.F. Ogston D. (1975) The effect of vitamin C supplements on serum cholesterol, coagulation, fibrinolysis and platelet adhesiveness. Atherosclerosis, 21, 451-454.
[209] Yudkin J. (1972) Pure White and Deadly: the problem of sugar, Davis Poynter, London.
[210] Wal van der A.C. de Boer O.J. and Becker A.E. (2001) Pathology of acute coronary syndromes, Progress in Inflammation Research, (M. J. Parnham, Ed.), Birkhäuser Verlag, Basel, Switzerland.
[211] Weber C. Wolfgang E. Weber K. Weber P.C. (1996) Increased adhesiveness of isolated monocytes to epithelium is prevented by vitamin C intake in smokers, Circulation, 93, 1488-1492.
[212] Libby P. (2002) Atherosclerosis: the new view, Scientific American, May, 29-37.
[213] Virmani R. Kolodgie F.D. Burke A.P. and Farb A. (2001) Inflammation in coronary atherosclerosis - pathological aspects, Progress in Inflammation Research, (M. J. Parnham, Ed.), Birkhäuser Verlag, Basel, Switzerland.
[214] Schieffer B. and Drexler H. (2001) Role of interleukins in relation to the renin-angiotensin-system in atherosclerosis, Progress in Inflammation Research, (M. J. Parnham, Ed.), Birkhäuser Verlag, Basel, Switzerland.

[215] Schmitz G. and Torzewski M. (2001) Atherosclerosis: an inflammatory disease, Progress in Inflammation Research, (M. J. Parnham, Ed.), Birkhäuser Verlag, Basel, Switzerland.

[216] Morrow D.A. and Ridker P.M. (2001) C-reactive protein - a prognostic marker of inflammation in atherosclerosis, Progress in Inflammation Research, (M. J. Parnham, Ed.), Birkhäuser Verlag, Basel, Switzerland.

[217] Rader D.J. (2000) Inflammatory markers of coronary risk, N Engl J Med, Oct 19, 343(16), 1179-1182.

[218] Romeo F. Clementi F. Saldeen T. Mehtas J.L. (2001) Role of infection in atherosclerosis and precipitation of acute cardiac events, Progress in Inflammation Research, (M. J. Parnham, Ed.), Birkhäuser Verlag, Basel, Switzerland.

[219] Shah P.K. (2000) Plaque disruption and thrombosis: potential role of inflammation and infection, Cardiol Rev, 8(1), 31-39.

[220] Gabay M.P. Jain R. (2002) Role of antibiotics for the prevention of cardiovascular disease, Ann Pharmacotherapy, 36(10), 1629-1636.

[221] Scannapieco F.A. Genco R.J. (1999) Association of periodontal infections with atherosclerotic and pulmonary diseases, J Periodontal Res, 34(7), 340-345.

[222] Shovman O. Levy Y. Gilburd B. Shoenfeld Y. (2001) Anti-inflammatory and immunomodulatory properties of statins, Rheuma21st, Online Journal.

[223] Blake G.J. Ridker P.M. (2000) Are statins anti-inflammatory? Curr Control Trials Cardiovasc Med, 1, 161-165.

[224] Sukhova G.K. Williams J.K. Libby P. (2002) Statins reduce inflammation in atheroma of nonhuman primates independent of effects on serum cholesterol, Arterioscler Thromb Vasc Biol, 22(9), 1452-1458.

[225] Aikawa M. Sugiyama S. Hill C.C. Voglic S.J. Rabkin E. Fukumoto Y. Schoen F.J. Witztum J.L. Libby P. (2002) Lipid lowering reduces oxidative stress and endothelial cell activation in rabbit atheroma, Circulation, 106(11), 1390-1396.

[226] Souza H.P. Zweier J.L. (2001) Free radicals as mediators of inflammation in atherosclerosis, Progress in Inflammation Research, (M. J. Parnham, Ed.), Birkhäuser Verlag, Basel, Switzerland.

[227] Young S. McEneny J. (2001) Lipoprotein oxidation and atherosclerosis, Biochemical Society Transactions, 29, 2, 358-361.

[228] Napoli C. Lerman L.O. (2001) Involvement of Oxidation-Sensitive Mechanisms in the Cardiovascular Effects of Hypercholesterolaemia, Mayo Clin Proc, 76, 619-631.

[229] Frei B. Stocker R. Ames B.N. (1988) Antioxidant defences and lipid peroxidation in blood plasma, Proc Natl Acad Sci USA, 85, 9748-9752.

[230] Jialal I. Vega G.L. Grundy S.M. (1990) Physiological levels of ascorbate inhibit the oxidative modification of low density lipoprotein, Atherosclerosis, 82, 185-191.

[231] Frei B. England L. Ames B.N.(1989) Ascorbate is the outstanding antioxidant in human blood plasma, Proc Natl Acad Sci USA, 86, 6377-6381.

[232] Kooyenga, D.K. Geller M. Watkins T.R. Bierenbaum, M.L. (1997) Antioxidant-induced regression of carotid stenosis over three-years, Proceedings of the 16th International Congress of Nutrition, Montreal, Canada.

[233] Witztum J.L Steinberg D. (1991) Role of oxidised low density protein in atherogenesis, J. Clin Invest, 88, 1785-1792.

[234] Jialal I. Devaraj S. (1996) The role of oxidised low density lipoprotein in atherogenesis, J Nutrit, 126, 1053S-1057S.

[235] Halliwell B. (1993) The role of oxygen radicals in human disease, with particular reference to the vascular system, Haemostasis, 23 Suppl 1, 118-126.

[236] Topper J.N. Cai J. Falb D. Gimbrone M.A. (1996) Identification of vascular endothelial genes differentially responsive to fluid mechanical stimuli: cyclooxygenase-2, manganese superoxide dismutase, and endothelial cell nitric oxide synthase are selectively up-regulated by steady laminar shear stress, Proc Natl Acad Sci U S A., 93(19), 10417-10422.

[237] Loscalzo J. (1996) The oxidant stress of hyperhomocyst(e)inemia, J Clin Invest, 98, 5.

[238] Dudman N.P. Wilcken D.E. Stocker R. (1993) Circulating lipid hydroperoxide levels in human hyperhomocysteinemia, Relevance to development of arteriosclerosis, Arterioscler Thromb, 13(4), 512-516.

[239] Navab M. Berliner J.A. Watson A.D. Hama S.Y. Territo M.C. Lusis A.J. Shih D.M. Van Lenten B.J. Frank J.S. Demer L.L. Edwards P.A. Fogelman A.M. (1996) The Yin and Yang of oxidation in the development of the fatty streak. A review based on the 1994 George Lyman Duff Memorial Lecture, Arterioscler Thromb Vasc Biol, 16(7), 831-842.

[240] Clare K. Hardwick S.J. Carpenter K.L. Weeratunge N. Mitchinson M.J. (1995) Toxicity of oxysterols to human monocyte-macrophages, Atherosclerosis, 118(1), 67-75.

[241] Parhami F. Morrow A.D. Balucan J. Leitinger N. Watson A.D. Tintut Y. Berliner J.A. Demer L.L. (1997) Lipid oxidation products have opposite effects on calcifying vascular cell and bone cell differentiation. A possible explanation for the paradox of arterial calcification in osteoporotic patients, Arterioscler Thromb Vasc Biol, 17(4), 680-687.

[242] Halpert I. Sires U.I. Roby J.D. Potter-Perigo S. Wight T.N. Shapiro S.D. Welgus H.G. Wickline S.A. Parks W.C. (1996) Matrilysin is expressed by lipid-laden macrophages at sites of potential rupture in atherosclerotic lesions and localizes to areas of versican deposition, a proteoglycan substrate for the enzyme, Proc Natl Acad Sci U S A, 93(18), 9748-9753.

[243] Parthasarathy S. Steinberg D. Witztum J.L. (1992) The role of oxidized low-density lipoproteins in the pathogenesis of atherosclerosis, Annu Rev Med, 43, 219-225.

[244] Han D.K. Haudenschild C.C. Hong M.K. Tinkle B.T. Leon M.B. Liau G. (1995) Evidence for apoptosis in human atherogenesis and in a rat vascular injury model, Am J Pathol, 147(2), 267-277.

[245] Darley-Usmar V. Halliwell B. (1996) Blood radicals, Pharm Res, 13, 649-662.

[246] Esterbauer H. Wag G. Puhl H. (1993) Lipid peroxidation and its role in atherosclerosis, Br Med Bull, 49(3), 566-576.

[247] Esterbauer H. Gebicki J. Puhl H. Jurgens G. (1992) The role of lipid peroxidation and antioxidants in oxidative modification of LDL, Free Radic Biol Med, 13(4), 341-390.

[248] Garner B. Witting P.K. Waldeck A.R. Christison J.K. Raftery M. Stocker R. (1998) Oxidation of high density lipoproteins. I. Formation of methionine sulfoxide in apolipoproteins AI and AII is an early event that accompanies lipid peroxidation and can be enhanced by alpha-tocopherol, J Biol Chem, 13, 273(11), 6080-6087.

[249] Garner B. Waldeck A.R. Witting P.K. Rye K.A. Stocker R. (1998) Oxidation of high density lipoproteins II, Evidence for direct reduction of lipid hydroperoxides by methionine residues of apolipoproteins AI and AII, J Biol Chem, 13, 273(11), 6088-6095.

[250] Bowry V.W. Mohr D. Cleary J. Stocker R. (1995) Prevention of tocopherol-mediated peroxidation in ubiquinol-10-free human low density lipoprotein, J Biol Chem, 17, 270(11), 5756-5763.

[251] Martin A. Frei B. (1997) Both intracellular and extracellular vitamin C inhibit atherogenic modification of LDL by human vascular endothelial cells, Arterioscler Thromb Vasc Biol, 17(8), 1583-1590.

[252] Gokce N, Frei B. (1996) Basic research in antioxidant inhibition of steps in atherogenesis, J Cardiovasc Risk, 3(4), 352-357.

[253] Jialal I. Fuller C.J. (1995) Effect of vitamin E, vitamin C and beta-carotene on LDL oxidation and atherosclerosis, Can J Cardiol, 11 Suppl G, 97G-103G.

[254] Retsky K.L. Frei B. (1995) Vitamin C prevents metal ion-dependent initiation and propagation of lipid peroxidation in human low-density lipoprotein, Biochim Biophys Acta, 3, 1257(3), 279-287.

[255] Mezzetti A. Lapenna D. Pierdomenico S.D. Calafiore A.M. Costantini F. Riario-Sforza G. Imbastaro T. Neri M. Cuccurullo F..(1995) Vitamins E, C and lipid peroxidation in plasma and arterial tissue of smokers and non-smokers, Atherosclerosis, 112(1), 91-99.

[256] Jialal I. Grundy S.M. (1991) Preservation of endogenous antioxidants in low density lipoprotein by ascorbate but not probucol during oxidative modification, J. Clin Invest, 87, 597-601.

[257] Scaccini C. Jialal I. (1994) LDL modification by activated polymorphonuclear leukocytes: a cellular model of mild oxidative stress, Free Rad Biol Med, 16, 49-55.

[258] Carr A.C. Frei B. (1999) Towards a new recommended dietary allowance for vitamin C based on antioxidant and health effects in humans, Am J Clin Nutr, 69, 1086, 1087.

[259] Turley S.D. West C.E. Horton B.J. (1976) The role of ascorbic acid in the regulation of cholesterol metabolism and in the pathogenesis of atherosclerosis, Atherosclerosis, 24(1-2), 1-18.

[260] Price K.D. Price C.S. Reynolds R.D. (1996) Hyperglycemia-induced latent scurvy and atherosclerosis: the scorbutic-metaplasia hypothesis, Med Hypotheses, 46(2), 119-129.

[261] Clemetson C.A. (1999) The key role of histamine in the development of atherosclerosis and coronary heart disease, Med Hypotheses, 52(1), 1-8.

[262] Schwarz T. Stoerk C.K. Renwick M. Millar R.G. Willis R. Paterson C. Sullivan M. (1999) Mineralisation of the coronary arteries in the dog, Abstract, American College of Veterinary Radiology, Annual Scientific Meeting, December 1-5, Chicago.

[263] Sako T. Takahashi T. Takehana K. Uchida E. Nakade T. Umemura T. Taniyama H. (2002) Chlamydial infection in canine atherosclerotic lesions, Atherosclerosis, 162(2), 253-259.

[264] Ginzinger D.G. Wilson J.E. Redenbach D. Lewis M.E.S. Clee S.M. Ashbourne Excoffon K.J.D. Rogers Q.R. Hayden M.R. McManus B.M. (1997) Diet-Induced Atherosclerosis in the Domestic Cat. Laboratory Investigation, November, 77(11).

[265] Chatterjee I.A. Majumder B. N. Subramanian N. (1975) Synthesis and some major functions of vitamin C in animals, Ann N. Y. Acad Sci, 258, 24-47.

[266] Attie A.D. Prescott M.F. (1988) The Spontaneously Hypercholesterolemic Pig as an Animal Model for Human Atherosclerosis, State of the Art, ILAR News, 30(4).

[267] Natarajan R Gerrity R.G. Gu J.L. Lanting L. Thomas L. Nadler J.L. (2002) Role of 12-lipoxygenase and oxidant stress in hyperglycaemia-induced acceleration of atherosclerosis in a diabetic pig model, Diabetologia, 45(1), 125-133.

[268] Vink-Nooteboom M. Schoemaker N.J. Kik M.J. Lumeij J.T. Wolvekamp W.T. (1998) Clinical diagnosis of aneurysm of the right coronary artery in a white cockatoo (Cacatua alba), J Small Anim Pract, 39(11), 533-537.

[269] Howerd A.N. (1976) The baboon in atherosclerosis research: comparison with other species and use in testing drugs affecting lipid metabolism, Adv Exp Med Biol, 67(00), 77-87.

[270] Prathap K. (1973) Spontaneous aortic lesions in wild adult Malaysian long-tailed monkeys (Macaca irus), J Pathol, 110(2), 135-143.

[271] Prathap K. (1975) Diet-induced aortic atherosclerosis in Malaysian long-tailed monkeys (Macaca irus), J Pathol, 115(3), 163-174.

[272] Linsay S. Chaikoff I.L. (1966) Naturally occurring atherosclerosis in non-human primates, J Atherosclerosis Research, 61, 36-61.

[273] Toien O. Drew K.L. Chao M.L. Rice M.E. (2001) Ascorbate dynamics and oxygen consumption during arousal from hibernation in Arctic ground squirrels, Am J Physiol Regulatory Integrative Comp Physiol, 281, 572–R583.

[274] Godin D.V. Dahlman D.M. (1993) Effects of hypercholesterolemia on tissue antioxidant status in two species differing in susceptibility to atherosclerosis, Res Commun Chem Pathol Pharmacol, 79(2), 151-166.

[275] Paterson J.C. (1940) Capillary rupture with intimal haemorrhage in the causation of Cerebral Vascular Lesions, Arch Path, 29, 345-354.

[276] Paterson J.C. (1941) Some factors in the causation of intimal hemorrhages and in the precipitation of coronary thrombi, Canad. M. A. J., Feb, 114-120.

[277] Bartley W, Krebs H.A. O'Brien J.R.P. (1953) Vitamin C requirement of human adults: a report by the vitamin C subcommittee of the Accessory Food Factors Committee and others, Medical Research Committee, London.

[278] Weindling P. (1996) Human guinea pigs and the ethics of experimentation: the BMJ's correspondent at the Nuremberg medical trial, British Medical Journal, 313, 1467-1470.

[279] Siow R.C. Sato H. Leake D.S. Ishii T. Bannai S. Mann G.E. (1999) Induction of antioxidant stress proteins in vascular endothelial and smooth muscle cells: protective action of vitamin C against atherogenic lipoproteins, Free Radic Res, 31(4), 309-318.

[280] Siow R.C. Richards J.P. Pedley K.C. Leake D.S. Mann G.E. (1999) Vitamin C protects human vascular smooth muscle cells against apoptosis induced by moderately oxidized LDL containing high levels of lipid hydroperoxides, Arterioscler Thromb Vasc Biol, 19(10), 2387-2394.

[281] Arroyo L.H. Lee R.T. (1999) Mechanisms of plaque rupture: mechanical and biologic interactions, Cardiovasc Res, 41(2), 369-375.

[282] Nakata Y, Maeda N. (2002) Vulnerable atherosclerotic plaque morphology in apoprotein E-deficient mice unable to make ascorbic Acid, Circulation Mar, 26, 105(12), 1485-1490.

[283] Willis G.C. (1953) An experimental study of the intimal ground substance in atherosclerosis, Canad. M. A. J. Vol 69, 17-22.

[284] De Nigris F. Lerman L.O. Ignarro W.S. Sica G. Lerman A. Palinski W. Ignarro L.J. Napoli C. (2003) Beneficial effects of antioxidants and L-arginine on oxidation-sensitive gene expression and endothelial NO synthase activity at sites of disturbed shear stress, Proc Natl Acad Sci U S A, epub ahead of print.

[285] Shmit E. (2003) Antioxidant Vitamins May Prevent Blood Vessel Blockage and Protect Against Cardiovascular Disease, UCLA News, January 15.

[286] Fernandez M.L. (2001) Guinea pigs as models for cholesterol and lipoprotein metabolism, J Nutr, 131(1), 10-20.

[287] Sulkin N.M. Sulkin D.F. (1975) Tissue changes induced by marginal vitamin C deficiency, Ann N Y Acad Sci, 258, 317-228.

[288] Rath M, Pauling L. (1990) Immunological evidence for the accumulation of lipoprotein(a) in the atherosclerotic lesion of the hypoascorbemic guinea pig, Proc Natl Acad Sci, Dec, 87(23), 9388-9390.

[289] Montano C.E. Fernandez M.L. McNamara D.J. (1998) Regulation of apolipoprotein B containing lipoproteins by vitamin C level and dietary fat saturation in guinea pigs, Metabolism, 47, 883-891.

[290] Satinder S. Sarkar A.K. Majumdar S. Chakravari R.N. (1987) Effects of ascorbic acid on the development of experimental atherosclerosis, Indian J. Med. Res, 86, 351-360.

[291] Yokota F. Igarashi Y. Suzue R. (1981) Hyperlipidemia in guinea pigs induced by ascorbic acid deficiency. Atherosclerosis, 38, 249-254.

[292] Ginter E. (1978) Marginal vitamin C deficiency, lipid metabolism and atherogenesis, Adv. Lipid Res, 16, 167-215.

[293] Liu J.F. Le Y.W. Vitamin C supplementation restores the impaired vitamin E status of guinea pigs fed oxidized frying oil, J. Nutr, 128, 16-122.

[294] Sharma P. Pramod J. Sharma P.K., Chaturvedi S.K. Kothari L.K. (1988) Effect of vitamin C administration on serum and aortic lipid profile of guinea pigs, Indian. J. Med. Res, 87, 283-287.

[295] Ravnskov U. (2002) A hypothesis out-of-date. The diet-heart idea, J Clin Epidemiol, 55(11), 1057-1063.

[296] Findlay G. (1921) A note on experimental scurvy in the rabbit and the effects of antenatal nutrition, The Journal of Pathology and Bacteriology, 24, 454-455.

[297] Finamore FJ, Feldman RP, Cosgrove GE. (1976) L-ascorbic acid, L-ascorbate 2-sulfate, and atherogenesis, Int J Vitam Nutr Res, 46(3), 275-285.

[298] Finamore F.J. Feldman R.P. Serrano L.J. Cosgrove G.E. (1977) L-ascorbate 2-sulfate and mobilization of cholesterol from plaque deposited in rabbit aortas, Int J Vitam Nutr Res, 47(1), 62-67.

[299] Hayashi E, Yamada J, Kunitomo M, Terada M, Sato M. (1978) Fundamental studies on physiological and pharmacological actions of L-ascorbate 2-sulfate, On the hypolipidemic and antiatherosclerotic effects of L-ascorbate 2-sulfate in rabbits, Jpn J Pharmacol, 28(1), 61-72..

[300] Mahfouz M.M. Kawano H. Kummerow F.A. (1997) Effect of cholesterol-rich diets with and without added vitamins E and C on the severity of atherosclerosis in rabbits, Am J Clin Nutr, 66(5), 1240-1249.

[301] Beetens J.R., Coene M.C. Veheyen A. Zonnekeyn L. Herman A.G. (1986) Vitamin C increases the prostacyclin production and decreases the vascular lesions in experimental atherosclerosis in rabbits, Prostaglandins, 32(3), 335-352.

[302] Verlangieri A.J. Hollis T.M. Mumma R.O. (1977) Effects of ascorbic acid and its 2-sulfate on rabbit aortic intimal thickening, Blood Vessels, 14(3), 157-174.

[303] Sun Y.P. Zhu B.Q. Sievers R.E. Norkus E.P. Parmley W.W. Deedwania P.C. (1998) Effects of antioxidant vitamins C and E on atherosclerosis in lipid-fed rabbits, Cardiology, 89(3), 189-194.

[304] Morel D.W., de la Llera-Moya M. Friday K.E. (1994) Treatment of cholesterol-fed rabbits with dietary vitamins E and C inhibits lipoprotein oxidation but not development of atherosclerosis, J Nutr, 124(11), 2123-2130.

[305] Braesen J.H. Beisiegel U. Niendorf A. (1995) Probucol inhibits not only the progression of atherosclerotic disease, but causes a different composition of atherosclerotic lesions in WHHL-rabbits, Virchows Arch, 426(2), 179-188.

[306] Lee J.Y. Hanna A.N. Lott J.A. Sharma H.M. (1996) The antioxidant and antiatherogenic effects of MAK-4 in WHHL rabbits, J Altern Complement Med, 2(4), 463-478.

[307] Schwenke D.C. Behr S.R. (1998) Vitamin E combined with selenium inhibits atherosclerosis in hypercholesterolemic rabbits independently of effects on plasma cholesterol concentrations, Circ Res, 83(4), 366-377.

[308] Altman R.F. Schaeffer G.M. Salles C.A. Ramos de Souza A.S. Cotias P.M. (1980) Phospholipids associated with vitamin C in experimental atherosclerosis, Arzneimittelforschung, 30(4), 627-630.

[309] Tsimikas S. Shortal B.P. Witztum J.L. Palinski W. (2000) In vivo uptake of radiolabeled MDA2, an oxidation-specific monoclonal antibody, provides an accurate measure of atherosclerotic lesions rich in oxidized LDL and is highly sensitive to their regression, Arterioscler Thromb Vasc Biol, 20(3), 689-697.

[310] Maeda N. Hagihara H. Nakata Y. Hiller S. Wilder J. Reddick R. (2000) Aortic wall damage in mice unable to synthesize ascorbic acid, 97(2), 841-846.

[311] Crawford R.S. Kirk E.A. Rosenfeld M.E. LeBoeuf R.C. Chait A. (1998) Dietary antioxidants inhibit development of fatty streak lesions in the LDL receptor-deficient mouse Arterioscler Thromb Vasc Biol, 18(9), 1506-1513.

[312] Willis G.C. (1957) The reversibility of atherosclerosis, Canad. M. A. J., 77, 106-109.

[313] Willis G.C., Light A.W., Cow W.S. (1954) Serial arteriography in atherosclerosis, Canad. M. A. J., 71, 562-568.

[314] Rath M. Niedzwiecki A. (1996) Nutritional Supplement Program Halts Progression of Early Coronary Atherosclerosis Documented by Ultrafast Computed Tomography, Journal of Applied Nutrition, 48, 68-78.

[315] Tomoda H; Yoshitake M; Morimoto K; Aoki N (1996) Possible prevention of postangioplasty restenosis by ascorbic acid, Am J Cardiol (US), Dec, 78 (11), 1284-1286.

[316] Fang JC, Kinlay S, Beltrame J, Hikiti H, Wainstein M, Behrendt D, Suh J, Frei B, Mudge GH, Selwyn AP, Ganz P. (2002) Effect of vitamins C and E on progression of transplant-associated arteriosclerosis: a randomised trial, Lancet, 30, 359(9312), 1108-1113.

[317] Langlois M. Duprez D. Delanghe J. De Buyzere M. Clement D.L. (2001) Serum vitamin C concentration is low in peripheral arterial disease and is associated with inflammation and severity of atherosclerosis, Circulation, 103(14), 1863-1868.

[318] Gackowski D, Kruszewski M, Jawien A, Ciecierski M, Olinski R. (2001) Further evidence that oxidative stress may be a risk factor responsible for the development of atherosclerosis, Free Radic Biol Med, 31(4), 542-547.

[319] Valkonen M.M. Kuusi T. (2000) Vitamin C prevents the acute atherogenic effects of passive smoking, Free Radic Biol Med, 28(3), 428-436.

[320] Frei B. (1999) On the role of vitamin C and other antioxidants in atherogenesis and vascular dysfunction, Proc Soc Exp Biol Med, 222(3), 196-204.

[321] Wilkinson I.B. Megson I.L. MacCallum H. Sogo N. Cockcroft J.R. Webb D.J. (1999) Oral vitamin C reduces arterial stiffness and platelet aggregation in humans, J Cardiovasc Pharmacol, 34(5), 690-693.

[322] Frei B. (1995) Cardiovascular disease and nutrient antioxidants: role of low-density lipoprotein oxidation, Crit Rev Food Sci Nutr, 35(1-2), 83-98.

[323] Eichholzer M, Stahelin HB, Gey KF (1992) Inverse correlation between essential antioxidants in plasma and subsequent risk to develop cancer, ischemic heart disease and stroke respectively: 12-year follow-up of the Prospective Basel Study, (1992) EXS, 62, 398-410.

[324] Jacob R.A. (1998) Vitamin C nutriture and risk of atherosclerotic heart disease, Nutr Rev, 56(11), 334-337.

[325] Ascherio A. Rimm E.B. Hernan M.A. Giovannucci E. Kawachi I. Stampfer M.J. Willett W.C. (1999) Relation of consumption of vitamin E, vitamin C, and carotenoids to risk for stroke among men in the United States, Ann Intern Med, 130(12), 963-970.

[326] Mayer-Davis EJ, Monaco JH, Marshall JA, Rushing J, Juhaeri (1997) Vitamin C intake and cardiovascular disease risk factors in persons with non-insulin-dependent diabetes mellitus. From the Insulin Resistance Atherosclerosis Study and the San Luis Valley Diabetes Study, Prev Med, 26(3), 277-283.

[327] Salonen J.T. Nyyssonen K. Salonen R. Lakka H.M. Kaikkonen J. Porkkala-Sarataho E. Voutilainen S. Lakka T.A. Rissanen T. Leskinen L. Tuomainen T.P. Valkonen V.P. Ristonmaa U. Poulsen H.E. (2000) Antioxidant Supplementation in Atherosclerosis Prevention (ASAP) study: a randomized trial of the effect of vitamins E and C on 3-year progression of carotid atherosclerosis, J Intern Med, 248(5), 377-386.

[328] Salonen RM, Nyyssonen K, Kaikkonen J, Porkkala-Sarataho E, Voutilainen S, Rissanen TH, Tuomainen TP, Valkonen VP, Ristonmaa U, Lakka HM, Vanharanta M, Salonen JT, Poulsen HE (2003) Six-year effect of combined vitamin C and E supplementation on atherosclerotic progression: the Antioxidant Supplementation in Atherosclerosis Prevention (ASAP) Study, Circulation, 107(7), 947-953.

[329] Gale C.R. Ashurst H.E. Powers H.J. Martyn C.N. (2001) Antioxidant vitamin status and carotid atherosclerosis in the elderly, Am J Clin Nutr, 74(3), 402-408.

[330] Lynch S.M. Gaziano J.M. Frei B. (1996) Ascorbic acid and atherosclerotic cardiovascular disease, Subcell Biochem (England), 25, 331-367.

[331] Ness AR; Powles JW; Khaw KT (1996) Vitamin C and cardiovascular disease: a systematic review, J Cardiovasc Risk, 3 (6), 513-521.

[332] Willis G.C. and Fishman S. (1955) Ascorbic acid content of human arterial tissue, Canad. M.A.J., April 1, Vol 72, 500-503.

[333] Suarna C. Dean R.T. May J. Stocker R. (1995) Human atherosclerotic plaque contains both oxidized lipids and relatively large amounts of alpha-tocopherol and ascorbate, Arterioscler Thromb Vasc Biol, 15(10), 1616-1624.

[334] O'Brien K.D. Alpers C.E. Hokanson J.E. Wang S. Chait A. (1996) Oxidation-specific epitopes in human coronary atherosclerosis are not limited to oxidized low-density lipoprotein, Circulation, 94(6), 1216-1225.

[335] Westhuyzen J. (1997) The oxidation hypothesis of atherosclerosis: an update, Ann Clin Lab Sci, 27(1), 1-10.

[336] Reaven P.D. Witztum J.L. (1996) Oxidized low density lipoproteins in atherogenesis: role of dietary modification, Annu Rev Nutr, 16, 51-71.

[337] Meagher E. Rader D.J. (2001) Antioxidant therapy and atherosclerosis: animal and human studies, Trends Cardiovasc Med, 11(3-4), 162-165.

[338] Rath M, Pauling L. (1991) Solution to the puzzle of human cardiovascular disease: Its primary cause is ascorbate deficiency, leading to the deposition of lipoprotein (a) and fibrinogen/fibrin in the vascular wall, Journal of Orthomolecular Medicine, 6, 125-134.

[339] Rath M Pauling L. (1989) Unified Theory of Human Cardiovascular Disease Leading the Way to the Abolition of this Disease as a Cause for Human Mortality, Arteriosclerosis, 9, 579-592.

[340] Rath M and Pauling L. (1991) Apoprotein(a) is an Adhesive Protein, Journal of Orthomolecular Medicine, 6, 139-143.

[341] Pauling L. Rath M. (1994) Prevention and treatment of occlusive cardiovascular disease with ascorbate and substances that inhibit the binding of lipoprotein (A), US patent 5, 278, 189.

[342] Pauling L. Rath M. (1993) Use of ascorbate and tranexamic acid solution for organ and blood vessel treatment prior to transplantation, US patent #5, 230, 996.

[343] Niendorf A. Rath M. Wolf K. Peters S. Arps H. Beisiegel U and Dietel M. (1990) Morphological detection and quantification of lipoprotein(a) deposition in atheromatous lesions of human aorta and coronary arteries, Virchows Archives of Pathological Anatomy, 417, 105-111.

[344] Rath M., Niendorf A., Reblin T., Dietel M., Krebber H.J., and Beisiegel U. (1989) Detection and Quantification of Lipoprotein (a) in the Arterial Wall of 107 Coronary Bypass Patients Arteriosclerosis, Vol 9, No 5, Sept/Oct.

[345] Beisiegel U. Niendorf A. Wolf K. Reblin T. Rath M. (1990) Lipoprotein (a) in the arterial wall, European Heart Journal, 11 (Suppl. E), 174-183.

[346] Nachman RL, Gavish D, Azrolan N, Clarkson TB. (1991) Lipoprotein(a) in diet-induced atherosclerosis in nonhuman primates. Arterioscler Thromb, 11(1), 32-38.

[347] Rath M. (1989) Reducing the risk for cardiovascular disease with nutritional supplements, Arteriosclerosis, 9, 579-592.

[348] Scanu A.M, Lawn R.M. Berg K. (1991) Lipoprotein(a) and atherosclerosis, Ann Intern Med, Aug, 1, 115(3), 209-218.

[349] Caplice N.M Panetta C, Peterson T.E. Kleppe L.S, Mueske C.S. Kostner G.M. Broze G.J. Simari R.D. (2001) Lipoprotein (a) binds and inactivates tissue factor pathway inhibitor: a novel link between lipoproteins and thrombosis, Blood, Nov, 15, 98(10), 2980-2987.

[350] Poon M. Zhang X. Dunsky K. Taubman M.B. Harpel P.C. (1997) Apoprotein(a) is a human vascular endothelial cell agonist: studies on the induction in endothelial cells of monocyte chemotactic factor activity, Clin Genetics, Nov, 52(5), 308-313.

[351] Scanu A.M. (1992) Lipoprotein(a): its inheritance and molecular basis of its atherothrombotic role, Mol Cell Biochem, Aug, 18, 113(2), 127-131.

[352] Reblin T. Meyer N. Labeur C. Henne-Bruns D. Beisiegel U. (1995) Extraction of lipoprotein(a), apo B, and apo E from fresh human arterial wall and atherosclerotic plaques, Atherosclerosis, Mar, 113(2), 179-188.

[353] Hoff H.F. O'Neil J. Yashiro A. (1993) Partial characterization of lipoproteins containing apo[a] in human atherosclerotic lesions, J. Lipid Res, May, 34(5), 789-798.

[354] Paultre F, Tuck CH, Boden-Albala B, Kargman DE, Todd E, Jones J, Paik MC, Sacco RL, Berglund L. (2002) Relation of Apo(a) size to carotid atherosclerosis in an elderly multiethnic population, Arterioscler Thromb Vasc Biol, Jan, 22(1), 141-146

[355] Yamamoto M. Egusa G. Yamakido M.(1997) Carotid atherosclerosis and serum lipoprotein(a) concentrations in patients with NIDDM, Diabetes Care, May, 20(5), 829-831.

[356] Kronenberg F, Kathrein H, Konig P, Neyer U, Sturm W, Lhotta K, Grochenig E, Utermann G, Dieplinger H. (1994) Apoprotein(a) phenotypes predict the risk for carotid atherosclerosis in patients with end-stage renal disease, Arterioscler Thromb, Sep, 14(9), 1405-1411.

[357] Pepin J.M. O'Neil J.A. Hoff H.F. (1991) Quantification of apo[a] and apoB in human atherosclerotic lesions, J Lipid Res, 32(2), 317-327.

[358] Stein J.H. Rosenson R.S. (1997) Lipoprotein Lp(a) excess and coronary heart disease, Arch Intern Med, 157(11), 1170-1176.

[359] Harris E.D. (1997) Lipoprotein[a]: a predictor of atherosclerotic disease, Nutr Rev, 55(3), 61-64.

[360] Miyao M. Araki A. Hattori A. Miyachi T. Inoue J. Horiuchi T. Nakamura T. Ueda S. Nakahara K. Matsushita S. Ito H. (1997) Lipoprotein(a) is not associated with coronary

heart disease in the elderly: cross-sectional data from the Dubbo study, Nippon Ronen Igakkai Zasshi, 34(3), 185-191

[361] Kubo M. Takami S. Matsuzawa Y. (1995) Contribution of Lp(a) to the occurrence of vascular diseases: correlation of several risk factors including diabetes mellitus, J Atheroscler Thromb, 2 Suppl 1, S22-S25.

[362] Simons L, Friedlander Y, Simons J, McCallum J. (1993) Atherosclerosis, 99(1), 87-95.

[363] Gazzaruso C, Buscaglia P, Garzaniti A, Falcone C, Mariotti S, Savino S, Bonetti G, Finardi G, Geroldi D. (1997) Association of lipoprotein(a) levels and apoprotein(a) phenotypes with coronary heart disease in patients with essential hypertension, J Hypertens, Mar, 15(3), 227-235.

[364] Marcovina SM, Hegele RA, Koschinsky ML. (1999) Lipoprotein(a) and coronary heart disease risk, Curr Cardiol Rep, Jul;1(2), 105-11.

[365] Nascetti S. D'Addato S. Pascarelli N. Sangiorgi Z. Grippo M.C. Gaddi A. (1996) Cardiovascular disease and Lp(a) in the adult population and in the elderly: the Brisighella study, Riv Eur Sci Med Farmacol, 18(5-6), 205-212.

[366] Kostner K.M. Huber K. Stefenelli T. Rinner H. Maurer G. (1997) Urinary apo(a) discriminates coronary artery disease patients from controls, Atherosclerosis, 129(1), 103-110.

[367] Nielsen L.B. Gronholdt M.L. Schroeder T.V. Stender S. Nordestgaard B.G. (1997) In vivo transfer of lipoprotein(a) into human atherosclerotic carotid arterial intima, Arterioscler Thromb Vasc Biol, 17(5), 905-911.

[368] Rath M. and Pauling L. (1990) Lipoprotein (a) is a surrogate for ascorbate, Proc Natl Acad Sci, 87, 6204-6207.

[369] Bostom A.G. Hume A.L. Eaton C.B. Laurino J.P. Yanek L.R. Regan M.S. McQuade W.H. Craig W.Y. Perrone G. Jacques P.F. (1995) The effect of high-dose ascorbate supplementation on plasma lipoprotein(a) levels in patients with premature coronary heart disease, Pharmacotherapy, 15(4), 458-464.

[370] Jenner J.L. Jacques P.F. Seman L.J. Schaefer E.J. (2000) Ascorbic acid supplementation does not lower plasma lipoprotein(a) concentrations, Atherosclerosis, 151(2), 541-544.

[371] Klezovitch O. Edelstein C. Scanu A.M. (1996) Evidence that the fibrinogen binding domain of Apo(a) is outside the lysine binding site of kringle IV-10: a study involving naturally occurring lysine binding defective lipoprotein(a) phenotypes, J Clin Invest, 98(1), 185-191.

[372] Angles-Cano E. (2000) High antifibrinolytic activity of lipoprotein(a) containing small apolipoprotein(a) isoforms, Circulation, 19, 102(25), E184-185.

[373] Angles-Cano E. de la Pena Diaz A. Loyau S. (2001) Inhibition of fibrinolysis by lipoprotein(a), Ann N Y Acad Sci, USA, 936, 261-275.

[374] Edelberg J.M. Pizzo S.V. (1995) Lipoprotein (a) in the regulation of fibrinolysis, J Atheroscler Thromb, 2 Suppl 1, S5-S7.

[375] Tousoulis D. Davies G. Ambrose J. Tentolouris C. Stefanadis C. Toutouzas P. (2002) Effects of lipids on thrombotic mechanisms in atherosclerosis, Int J Cardiol, 86(2-3), 239-247.

[376] Drouet L. Bal Dit Sollier C. (2002) Fibrinogen: factor and marker of cardiovascular risk, J Mal Vasc, 27(3), 143-156.

[377] Angles-Cano E. Rojas G. (2002) Apolipoprotein(a): structure-function relationship at the lysine-binding site and plasminogen activator cleavage site, Biol Chem, 383(1), 93-99.

[378] Kang C. Dominguez M. Loyau S. Miyata T. Durlach V. Angles-Cano E. (2002) Lp(a) particles mold fibrin-binding properties of apo(a) in size-dependent manner: a study with different-length recombinant apo(a), native Lp(a), and monoclonal antibody, Arterioscler Thromb Vasc Biol, 2(7), 1232-8.

[379] Boonmark N.W. Lou X.J. Yang Z.J. Schwartz K. Zhang J.L. Rubin E.M. Lawn R. (1997) Modification of apolipoprotein(a) lysine binding site reduces atherosclerosis in transgenic mice, J Clin Invest, 100(3), 558-564.

[380] Higazi A.A. Lavi E. Bdeir K. Ulrich A.M. Jamieson D.G. Rader D.J. Usher D.C. Kane W. Ganz T. Cines D.B. (1997) Defensin stimulates the binding of lipoprotein (a) to human vascular endothelial and smooth muscle cells, Blood, 89(12), 4290-4298.

[381] Eaton D.L. Fless G.M. Kohr W.J. McLean J.W. Xu Q.T. Miller C.G. Lawn R.M. Scanu A.M. (1987) Partial amino acid sequence of apolipoprotein(a) shows that it is homologous to plasminogen, Proc Natl Acad Sci U S A, 84(10), 3224-3228.

[382] Kratzin H. Armstrong V.W. Niehaus M. Hilschmann N. Seidel D. (1987) Structural relationship of an apolipoprotein (a) phenotype (570 kDa) to plasminogen: homologous kringle domains are linked by carbohydrate-rich regions, Biol Chem Hoppe Seyler, 368(12), 1533-1544.

[383] McLean J.W. Tomlinson J.E. Kuang W.J. Eaton D.L. Chen E.Y. Fless G.M. Scanu A.M. Lawn R.M. (1987) cDNA sequence of human apolipoprotein(a) is homologous to plasminogen, Nature 330(6144), 132-137.

[384] Stubbs P. Seed M. Moseley D. O'Connor B. Collinson P. Noble M. (1997) A prospective study of the role of lipoprotein(a) in the pathogenesis of unstable angina, Eur Heart J, 18(4), 603-607.

[385] Rath M, Pauling L. (1992) Plasmin-induced proteolysis and the role of apoprotein(a), lysine, and synthetic lysine analogs, Journal of Orthomolecular Medicine, 7, 17-23.

[386] Soulat T. Loyau S. Baudouin V. Durlach V. Gillery P. Garnotel R. Loirat C. Angles-Cano E. (1999) Evidence that modifications of Lp(a) in vivo inhibit plasmin formation on fibrin-a study with individual plasmas presenting natural variations of Lp(a), Thromb Haemost, 82(1), 121-127.

[387] Angles-Cano E. (1994) Overview on fibrinolysis: plasminogen activation pathways on fibrin and cell surfaces, Chem Phys Lipids, 67-68, 353-62.

[388] Tsuchihashi K. Minari O. (1983) Lysine residues located on the surface of human plasma high-density lipoprotein particles, Biochim Biophys Acta, 752(1), 10-18.

[389] Fong B.S. Rodrigues P.O. Angel A.J. (1984) Characterization of low density lipoprotein binding to human adipocytes and adipocyte membranes, J Biol Chem, 259(16), 10168-10174.

[390] van Willigen G. Gorter G. Akkerman J.W. (1994) LDLs increase the exposure of fibrinogen binding sites on platelets and secretion of dense granules, Arterioscler Thromb, 14(1), 41-46.

[391] Shute E. (1964) The current status of alpha tocopherol in cardiovascular disease, In Vitamin E, your key to a healthy heart, by Herbert Bailey, ARC Books, New York.

[392] Stephens, N.G.; Parsons, A.; Schofield, P.M.; Kelly, E; Cheeseman, K.; Mitchinson, M.J. and Brown, M J. (1996) Randomised Controlled Trial of Vitamin E in Patients with Coronary Disease: Cambridge Heart Antioxidant Study (CHAOS) Lancet, 349, 781-786.

[393] Mitchinson, M.J. Stephens N.G. Parsons A. Bligh E. Schofield P.M. Brown M.J. (1999) Mortality in the CHAOS Trial, Lancet, 353, 381- 382.

[394] Losonczy KG; Harris TB; Havlik RJ (1996) Epidemiology, Vitamin E and vitamin C supplement use and risk of all-cause and coronary heart disease mortality in older persons: the Established Populations for Epidemiologic Studies of the Elderly, Am J Clin Nutr (US), Aug, 64 (2) 190-196.

[395] Rimm E.B. Stampfer M.J. Ascherio A. Giovannucci E. Colditz G.A. Willett W.C. (1993) Vitamin E consumption and the risk of coronary heart disease in men, N Engl J Med, 328(20), 1450-1456.

[396] Stampfer M.J. Hennekens C.H. Manson J.E. Colditz G.A. Rosner B. Willett W.C. (1993) Vitamin E consumption and the risk of coronary disease in women, N Engl J Med, 328(20), 1444-1449.

[397] Gey, K.F. and Puska, P (1989) Plasma Vitamin E and A Inversely Correlated to Mortality from Ischemic Heart Disease in Cross-Cultural epidemiology Ann. N. Y. Acad. Sci., 570, 268-282.

[398] Gey, K.E; Puska, P; Jordan, P and Moyer, U.K. (1991) Inverse Correlation Between Plasma Vitamin E and Mortality from Ischemic Heart Disease in Cross-Cultural Epidemiology, Amen J. Clin Nutr, 53, 3265-3345.

[399] Verlangieri, A.J. and Bush, M.K. (1992) Effects of d-alpha-tocopherol supplementation on experimentally induced primate atherosclerosis. J. Amer. Coll. Nutr. 11, 131-138.

[400] Passwater R.A. (1992) Reversing atherosclerosis: An interview with Dr. Anthony Verlangieri, Whole Foods, 15(9), 27-30.

[401] Hodis, H.N.; Mack, W.J.; LaBree, L.; Cashin-Hemphill, L.; Sevanian, A.; Johnson, R. and Azen, S.P. (1995) Serial coronary angiographic evidence that antioxidant vitamin intake reduces progression of coronary artery atherosclerosis, JAMA, 273(23), 1849-1854.

[402] DeMaio S.J. King S.B. Lembo N.J. Roubin G.S. Hearn J.A. Bhagavan H.N. Sgoutas D.S. (1992) Vitamin E supplementation, plasma lipids and incidence of restenosis after percutaneous transluminal coronary angioplasty (PTCA), J. Amer. Coll. Nutr, 11, 68-73.

[403] Kooyenga, D.K., Geller M., Watkins T.R. and Bierenbaum M.L. (1997) Antioxidant-induced regression of carotid stenosis over three-years, Proceedings of the 16th International Congress of Nutrition. Montreal.

[404] Tomeo A.C. Geller M. Watkins T.R. Gapor A. Bierenbaum M.L. (1995) Antioxidant effects of tocotrienols in patients with hyperlipidemia and carotid stenosis, Lipids, 30(12), 1179-1183.

[405] Black T.M. Wang P. Maeda N. Coleman R.A. (2000) Palm tocotrienols protect ApoE +/- mice from diet-induced atheroma formation, J Nutr, 130(10), 2420-2426.

[406] Qureshi A.A. Salser W.A. Parmar R. Emeson E.E. (2001) Novel tocotrienols of rice bran inhibit atherosclerotic lesions in C57BL/6 ApoE-deficient mice, J Nutr, 131(10), 2606-2618.

[407] Ismail N.M. Abdul Ghafar N. Jaarin K. Khine J.H. Top G.M. (2000) Vitamin E and factors affecting atherosclerosis in rabbits fed a cholesterol-rich diet, Int J Food Sci Nutr, 51, Suppl, S79-S94.

[408] Teoh M.K. Chong J.M. Mohamed J. Phang K.S. (1994) Protection by tocotrienols against hypercholesterolaemia and atheroma, Med J Malaysia, 49(3), 255-262.

[409] Micheletta F. Natoli S. Misuraca M. Sbarigia E. Diczfalusy U. Iuliano L. (2004) Vitamin E Supplementation in Patients With Carotid Atherosclerosis: Reversal of Altered Oxidative Stress Status in Plasma but not in Plaque, Arterioscler Thromb Vasc Biol, 24(1), 136-140.

[410] McCarty M.F. (1981) An expanded concept of "insurance" supplementation-broad-spectrum protection from cardiovascular disease, Med Hypotheses, 7(10), 1287-1302.

[411] Chan AC, Chow CK, Chiu D. (1999) Interaction of antioxidants and their implication in genetic anemia, Proc Soc Exp Biol Med, 222(3), 274-282.

[412] Shih J.C. (1983) Atherosclerosis in Japanese quail and the effect of lipoic acid, Fed Proc, 15, 42(8), 2494-2497.

[413] Shim K.F. and Vohra P. (1984) A review of the nutrition of Japanese quail, World's Poultry Sci J, 40, 261-274.

[414] Ramachandran V. Arscott, G.H. (1974) Minimum vitamin requirements and apparent vitamin interrelationships for growth in Japanese quail (Coturnix coturnix japonica), Poultry Sci, 53, 1969-1970.

[415] Vodoevich V.P. (1983) Effect of lipoic acid, biotin and pyridoxine on blood content of saturated and unsaturated fatty acids in ischemic heart disease and hypertension, Vopr Pitan, 5, 14-16.

[416] Achmad T.H. Rao G.S. (1992) Chemotaxis of human blood monocytes toward endothelin-1 and the influence of calcium channel blockers, Biochem Biophys Res Commun, 15, 189(2), 994-1000.

[417] Kunt T. Forst T. Wilhelm A. Tritschler H. Pfuetzner A. Harzer O. Engelbach M. Zschaebitz A. Stofft E. Beyer J. (1999) Alpha-lipoic acid reduces expression of vascular cell adhesion molecule-1 and endothelial adhesion of human monocytes after stimulation with advanced glycation end products, Clin Sci (Lond), 96(1), 75-82.

[418] Schleicher E.D. Wagner E. Nerlich A.G. (1997) Increased accumulation of the glycoxidation product N(epsilon)-(carboxymethyl)lysine in human tissues in diabetes and aging, J Clin Invest, 99(3), 457-468.

[419] Egan R.W. Gale P.H. Beveridge G.C. Phillips G.B. Marnett L.J. (1978) Radical scavenging as the mechanism for stimulation of prostaglandin cyclooxygenase and depression of inflammation by lipoic acid and sodium iodide, Prostaglandins, 16(6), 861-869.

[420] Kohler HB, Huchzermeyer B, Martin M, De Bruin A, Meier B, Nolte I. (2001) TNF-alpha dependent NF-kappa B activation in cultured canine keratinocytes is partly mediated by reactive oxygen species, Vet Dermatol, 12(3), 129-137.

[421] Mann J. (1999) The Elusive Magic Bullet, The Search for the Perfect Drug, Oxford University Press, Oxford, England.
[422] Levy T.E. (2002) Vitamin C, Infectious Disease and Toxins, Xlibris, Philadelphia.
[423] Jariwalla R.J Harakeh S. (1994) Ascorbic acid and AIDS: Strategic Functions and Therapeutic Possibilities, In Nutrition and AIDS, Ed Watson R.R., CRC Press, Boca Raton.
[424] Otani T. (1936) On the vitamin C therapy of pertussis, Klinische Wochenschrift, 15, 1884-1885.
[425] Ormerod M.B. Unkauf B.M. (1937) Ascorbic acid (vitamin C) treatment of whooping cough, The Canadian Medical Association Journal, 37, 2, 134-136.
[426] Ormerod M.B. Unkauf B.M. White F.D. (1937) A further report on the ascorbic acid treatment of whooping cough, The Canadian Medical Association Journal, 37, Sept 3, 268-272.
[427] Vermillion E. L. Stafford G.E. (1938) A preliminary report on the use of cvitamic acid in the treatment of whooping cough, The Journal of The Kansas Medical Society, XXXIX, Nov, 11, 469-479.
[428] Markwell M.W. (1947) Vitamin C in the Prevention of Colds, The Medical Journal of Australia, Dec 2, 26, 777-778.
[429] Klenner F.R (1948) Virus Pneumonia and Its Treatment With Vitamin C, Southern Medicine & Surgery, 110, Feb, 2, 36-38.
[430] Klenner F. R. (1949) The Treatment of Poliomyelitis and Other Virus Diseases with Vitamin C, Southern Medicine & Surgery, 111, 7, July, 209-214.
[431] Benedict F. Massell M.D. Warren J.E., Patterson P.R., Lehmus H.J. (1950) Antirheumatic activity of ascorbic acid in large doses: Preliminary Observations on Seven Patients with Rheumatic Fever, New England Journal of Medicine, 242, April, 1950, 16, 614-615.
[432] Klenner F.R. (1951) Massive Doses of Vitamin C and the Virus Diseases, Southern Medicine & Surgery, 103, 4, 101-107.
[433] McCormick W.J. (1951) Vitamin C in the prophylaxis and therapy of infectious diseases, Archives of Pediatrics, 68, 1, 1-9.
[434] McCormick W.J. (1952) Ascorbic acid as a chemotherapeutic agent, Archives of Pediatrics, 69, 4, 151-155.
[435] Klenner F. R. (1957) An insidious virus, Tri-State Med. J., June.
[436] Klenner F. (1974) Significance of High Daily Intake of Ascorbic Acid in Preventive Medicine, Journal of the International Academy of Preventive Medicine, 1, 1, 45-69.
[437] Stone I. (1974) The Healing Factor: Vitamin C Against Disease, Putnam, New York.
[438] Klenner F. Bartz F.H. (1969) The Key To Good Health: Vitamin C, Graphic Arts Research Foundation, Chicago.
[439] Schwerdt P.R, Schwerdt C.E. (1975) Effect of ascorbic acid on rhinovirus replication in WI-38 cells, Proc Soc Exp Biol Med, 148(4), 1237-1243.
[440] White L.A. Freeman C.Y. Forrester B.D. Chappell W.A. In vitro effect of ascorbic acid on infectivity of herpesviruses and paramyxoviruses, J Clin Microbiol, Oct, 24(4), 527-531.

[441] Sagripanti J.L. Routson L.B. Bonifacino A.C. Lytle C.D. (1997) Mechanism of copper-mediated inactivation of herpes simplex virus, Antimicrob Agents Chemother, Apr, 41(4), 812-817.

[442] Cinatl J. Cinatl J. Weber B. Rabenau H. Gumbel H.O. Chenot J.F. Scholz M. Encke A. Doerr H.W. (1995) In vitro inhibition of human cytomegalovirus replication in human foreskin fibroblasts and endothelial cells by ascorbic acid 2-phosphate, Antiviral Res, Aug, 27(4), 405-418.

[443] Chalmers T.C. (1975) Effects of ascorbic acid on the common cold; an evaluation of the evidence, American Journal of Medicine, 58, 532-536.

[444] Douglas R.M. Chalker E.B. Treacy B. (2001) Vitamin C for preventing and treating the common cold (Cochrane Review), In: The Cochrane Library, 3, Oxford: Update Software, CD000980.

[445] Hemila H. (1999) Vitamin C supplementation and common cold symptoms: factors affecting the magnitude of the benefit, Med Hypotheses, Feb, 52(2), 171-178.

[446] Hemila H. Douglas R.M. (1999) Vitamin C and acute respiratory infections, Int J Tuberc Lung Dis, 3(9), 756-761.

[447] Hemila H. (1992) Vitamin C and the common cold, Br J Nutr, Jan, 67(1), 3-16.

[448] Hemila H. (1994) Does vitamin C alleviate the symptoms of the common cold? - a review of current evidence, Scand J Infect Dis, 26(1), 1-6.

[449] Hemila H. J. Vitamin C. (1996) The placebo effect, and the common cold: a case study of how preconceptions influence the analysis of results, Clin Epidemiol, Oct, 49(10), 1079-1084.

[450] Hemila H. (1996) Vitamin C supplementation and common cold symptoms: problems with inaccurate reviews, Nutrition, Nov-Dec, 12(11-12), 804-809.

[451] Hemila H. (1997) Vitamin C intake and susceptibility to the common cold, Br J Nutr, Jan, 77(1), 59-72.

[452] Hemila H. (1997) Vitamin C supplementation and the common cold-was Linus Pauling right or wrong? Int J Vitam Nutr Res, 67(5), 329-335.

[453] Gorton H.C. Jarvis K. (1999) The effectiveness of vitamin C in preventing and relieving the symptoms of virus-induced respiratory infections, J Manipulative Physiol Ther, Oct, 22(8), 530-533.

[454] Van Straten M. Josling P. (2002) Preventing the common cold with a vitamin C supplement: a double-blind, placebo-controlled survey, Adv Ther, May-Jun, 19(3), 151-159.

[455] Audera C. Patulny R.V. Sander B.H. Douglas R.M. (2001) Mega-dose vitamin C in treatment of the common cold: a randomised controlled trial, Med J Aust, 175, 359-362.

[456] Toy M. (2001) Study rebuts 'myth' of vitamin C cold cure, TheAge.com, Monday 1 October.

[457] Anderson T.N. Suranyi B. Beaton G.W. (1974) The effect on winter illness of large doses of vitamin C, Can Med Assoc J, 111, 31-38.

[458] Karlowski T.R. Chalmers T.C. Frenkel L.D. Kapikian A.Z. Lewis T.L. Lynch J.M. (1975) Ascorbic acid for the common cold, A prophylactic and therapeutic trial, JAMA, 231, 1038-1042.

[459] Elwood P.C. Hughes S.J. St Leger A.S. (1977) A randomized controlled trial of the therapeutic effect of vitamin C in the common cold. Practitioner, 218, 133-137.

[460] Tyrrell D.A. Craig J.W. Meada T.W. White T. (1977) A trial of ascorbic acid in the treatment of the common cold, Br J Prev Soc Med, 31, 189-191.

[461] Oxford J.S. (2002) Influenza A: a threatening virus with two faces, Biologist, 49(2), 63-67.

[462] Kolata G. (1999) Flu, Pan Books, London, England.

[463] Magme R.V. (1963) Vitamin C in treatment of influenza, El Dia medico, 35, 1714-1715.

[464] Tantcheva L.P. Stoeva E.S. Galabov A.S. Braykova A.A. Savov V.M. Mileva M.M. (2003) Effect of vitamin E and vitamin C combination on experimental influenza virus infection, Methods Find Exp Clin Pharmacol, 25(4), 259-264.

[465] Jungeblut, C. W. (1937) Vitamin C Therapy and Prophylaxis in Experimental Poliomyelitis, J. Exper. Med., 65, 127.

[466] Jungeblut, C. W. (1937) Further observations on vitamin C. Therapy in Experimental Poliomyelitis, J Exper Med, 66, 459.

[467] Klenner R.F. (1953) The use of vitamin C as an antibiotic, The Journal of Applied Nutrition, 6, 274-278.

[468] Sabin, A. B. (1939) Vitamin C in Relation to Experimental Poliomyelitis, J Exper Med, 69, 507.

[469] Klenner F.R. (1949) The treatment of poliomyelitis and other virus diseases with vitamin C, Southern Medicine and Surgery, July, 209.

[470] Ramar S, Sivaramakrishnan V, Manoharan K. (1993) Scurvy-a forgotten disease, India. Arch Phys Med Rehabil, Jan, 74(1), 92-95.

[471] Greer E. (1955) Vitamin C In Acute Poliomyelitis, Medical Times, 83, Nov, 11, 1160-1161.

[472] Brighthope I. Fitzgerald P. (1987) The Aids Fighters, the role of vitamin C and other immunity building nutrients, Keats, Connecticut.

[473] Cathcart R.F (1983) Vitamin C function in AIDS, Current Opinion, Medical Tribune, July 13.

[474] Cathcart R. F. (1984) Vitamin C in the treatment of acquired immune deficiency syndrome (AIDS), Medical Hypotheses, 14, 423 433.

[475] Cathcart R.F. (1988) AIDS treatment using ascorbic acid, Townsend Letter for Doctors. AIDS treatment using ascorbic acid, Townsend Letter for Doctors.

[476] Jariwalla R.J. (1995) Micronutrient imbalance in HIV infection and AIDS: relevance to pathogenesis and therapy, Journal of Nutritional Medicine, 5, 297-306.

[477] Favier A. Sappey C. Leclerc P. Faure P. Micoud M. (1994) Antioxidant status and lipid peroxidation in patients infected with HIV, Chem Biol Interact, Jun, 91(2-3), 165-180.

[478] Harakeh S. Niedzwiecki A. Jariwalla R.J. (1994) Mechanistic aspects of ascorbate inhibition of human immunodeficiency virus, Chem Biol Interact, 91(2-3), 207-215.

[479] Harakeh S. Jariwalla R.J. (1994) Comparative analysis of ascorbate and AZT. Effects on HIV production in persistently infected cell lines, Journal of Nutritional Medicine, 4, 393-401.

[480] Harakeh S. Jariwalla R.J. (1995) Ascorbate effect on cytokine stimulation of HIV production, Nutrition, Sep-Oct, 11(5 Suppl), 684-687.

[481] Harakeh S. Jariwalla R.J. (1997) NF-kappa B-independent suppression of HIV expression by ascorbic acid, AIDS Res Hum Retroviruses, Feb, 13(3), 235-239.

[482] Hirano F. Tanaka H. Miura T. Hirano Y. Okamoto K. Makino Y. Makino I. (1998) Inhibition of NF-kappaB-dependent transcription of human immunodeficiency virus 1 promoter by a phosphodiester compound of vitamin C and vitamin E, EPC-K1, Immunopharmacology, Mar, 39(1), 31-38.

[483] Harakeh S. (1991) Comparative study of the anti-HIV activities of ascorbate and thiol reducing agents in chronically and acutely infected cells, Proc. Natural Academy of Sciences, 87, 1231S-1235S.

[484] Harakeh S. Jariwalla R.J. Pauling L. (1990) Suppression of human immunodeficiency virus replication by ascorbate in chronically and acutely infected cells, Proc Natl Acad Sci, USA, 87(18), 7245-7249.

[485] Aoki K. Nakashima H. Hattori T. Shiokawa D. Ni-imi E. Tanimoto Y. Maruta H. Uchiumi F. Kochi M. Yamamoto N. (1994) Sodium benzylideneascorbate induces apoptosis in HIV-replicating U1 cells, FEBS Lett, Aug, 29, 351(1), 105-108.

[486] Eylar E, Baez I, Navas J, Mercado C. (1996) Sustained levels of ascorbic acid are toxic and immunosuppressive for human T cells, P R Health Sci J, Mar, 15(1), 21-26.

[487] Cheung E.Mutahar R. Assefa F. Ververs M.T. Nasiri S.M. Borrel A. Salama P. (2002) An epidemic of scurvy in Afghanistan: assessment and response, Food Nutr Bull, 24(3), 247-255.

[488] Banic S. (1982) Immunostimulation by vitamin C, Int Journal of Vitamin and Nutrition Research, 23, 49-52.

[489] Nicol M. (1993) Vitamins and immunity, Allerg Immunol, 25(2), 70-73.

[490] Cathcart R.F. (1986) The vitamin treatment of allergies and the normally unprimed state of antibodies, Medical Hypotheses, 21(3), 307-321.

[491] Schwartz J. Weiss S.T. (1994) Relationship between dietary vitamin C intake and pulmonary function in first national health and nutrition survey (NHANES 1), American Journal of Clinical Nutrition, 59(1), 110-114.

[492] Mohensen V. (1987) effect of vitamin C on NO2-induced airway hyperresponsiveness in normal subjects: a randomised double-blind experiment, American review of Respiratory Disease, 136(6), 1408-1411.

[493] Devereux G. Seaton A. (2001) Why don't we give chest patients dietary advice? Thorax, 56 (Suppl 2), ii15-ii22.

[494] Null G. (2002) AIDS a Second Opinion, Seven Stories Press, New York.

[495] Yonemoto R.H. (1979) Vitamin C and the immune response in normal controls and in cancer patients, Medico Dialogo, 5, 23-30.

[496] Yonemoto R.H., Chretien P.B. and Fehniger T.F. (1976) Enhanced lymphocyte blastogenesis by oral ascorbic acid, Proc. Am. Assoc. Cancer Res, 17, 288.

[497] Anderson R. (1981) Ascorbate mediated stimulation of neutrophil motility and lymphocyte transformation by inhibition of the peroxidase H_2O_2 halide system in vitro and in vivo, American Journal of Clinical Nutrition, 34, 1906-1911.

[498] Anderson R. (1981) Assessment of oral ascorbate in 3 children with chronic granulomatus disease and defective neutrophil motility over a 2 year period, Clinical and Experimental Immunology, 43, 180-188.

[499] Anderson R. (1982) Effects of ascorbate on normal and abnormal leukocyte functions, in Vitamin C: New Clinical Applications in Immunology, Lipid Metabolism and Cancer, ed. A. Hank., Hans Huber, Bern, 23-34.

[500] Geber W.F. Lefkowitz S.S. Hung C.Y. (1975) Effect of ascorbic acid, sodium salicylate, and caffeine on the serum interferon level in response to viral infection, Pharmacology, 13(3), 228-233.

[501] Goodman S. (1991) Vitamin C, The Master Nutrient, Keats, Connecticut.

[502] Cameron E. Pauling L. (1993) Cancer and Vitamin C, Camino Books, Philadelphia.

[503] Hoffer A. Walker M. (1978) Putting It All Together: The New Orthomolecular Nutrition, Keats Publishing, Connecticut.

[504] Cooke R. (2001) Doctor Folkman's War, Century, London.

[505] Voght A. (1940) On the Vitamin C Treatment of Chronic Leukemias, Deutsche Medizinische Wochenschrift, April, 14, 369-372.

[506] McCormick W.J. (1954) Cancer: the preconditioning factor in pathogenesis. Arch Pediat, 71, 313–322.

[507] McCormick W.J. (1959) Cancer: a collagen disease, secondary to a nutritional deficiency? Arch Pediat, 76, 166–171.

[508] Cameron E, Rotman D. (1972) Ascorbic acid, cell proliferation, and cancer, Lancet, 1, 542.

[509] Cameron E. Pauling L. (1973) Ascorbic acid and the glycosaminoglycans, An orthomolecular approach to cancer and other diseases, Oncology, 27, 181–192.

[510] Cameron E. Campbell A. (1974) The orthomolecular treatment of cancer II. Clinical trial of high-dose ascorbic acid supplements in advanced human cancer, Chem Biol Interact, 9, 285–315.

[511] Cameron E. Pauling L. (1976) Supplemental ascorbate in the supportive treatment of cancer: prolongation of survival times in terminal human cancer, Proc Nat Acad Sci, USA, 73, 3685-3689.

[512] Cameron E, Pauling L. (1978) Supplemental ascorbate in the supportive treatment of cancer: reevaluation of prolongation of survival times in terminal human cancer, Proc Nat Acad Sci, 75, 4538-4542.

[513] Cameron E. (1980) Vitamin C for cancer, N Engl J Med, 302, 299.

[514] Cameron E. Campbell A. (1991) Innovation vs. quality control: an 'unpublishable' clinical trial of supplemental ascorbate in incurable cancer, Med Hypotheses, 36, 185–189.

[515] Campbell A. Jack T. Cameron E. (1991) Reticulum cell sarcoma: two complete 'spontaneous' regressions, in response to high-dose ascorbic acid therapy. A report on subsequent progress, Oncology, 48, 495–497.

[516] Stone I. (1972) The Healing Factor. "Vitamin C" against disease, Grosset and Dunlap Inc., New York.

[517] Stone I. (1976) The Genetics of Scurvy and the Cancer Problem, Orthomolecular Psychiatry, 5, 3, 183-190.

[518] Greer E. (1954) Alcoholic Cirrhosis: Complicated by Polycythemia Vera and Then Myelogenous Leukemia and Tolerance of Large Doses of Vitamin C, Med. Times 82, 765-768.

[519] Garattini S. Bertele V. (2002) Efficacy, safety, and cost of new anticancer drugs BMJ, 325, 269-271.

[520] Everson T.C. Cole W.H. (1966) Spontaneous Regression of Cancer, W.B. Saunders, Philadelphia.

[521] Boyd W. (1966) The Spontaneous Regression of Cancer, Charles C Thomas, Springfield.

[522] Murata A, Morishige F, Yamaguchi H. (1982) Prolongation of survival times of terminal cancer patients by administration of large doses of ascorbate, International Journal for Vitamin and Nutrition Research, Supplement, 23, 101-113.

[523] DeWys W.D. (1982) How to evaluate a new treatment for cancer, Your Patient and Cancer, 2(5), 31-36.

[524] Creagan E.T. Moertel C.G. O'Fallon J.R. Schutt A.J. O'Connell M.J. Rubin J. Frytak S. (1979) Failure of high-dose vitamin C (ascorbic acid) therapy to benefit patients with advanced cancer, A controlled trial, N Engl J Med, 301, 687–690.

[525] Moss M. (1992) The Cancer Industry, Equinox Press, New York.

[526] Pauling L. (1980) Vitamin C therapy and advanced cancer (letter). N Engl J Med, 302, 694.

[527] Moertel C.G. Creagan E.T. (1980) Vitamin C therapy and advanced cancer (letter), New England Journal of Medicine, 302, 694-695.

[528] Moertel C.G. Fleming T.R. Creagan E.T. Rubin J. O'Connell M.J. Ames M.M. (1985) High-dose vitamin C versus placebo in the treatment of patients with advanced cancer who have had no prior chemotherapy. A randomized double-blind comparison, N Engl J Med, 312, 137–141.

[529] Smith G.C. Pell J.P. (2003) Parachute use to prevent death and major trauma related to gravitational challenge: systematic review of randomised controlled trials, British medical Journal, 327(7429), 1459-1461.

[530] Herman Z. S. (2000) On the Understanding of the Hardin Jones-Pauling Biostatistical Theory of Survival Analysis for Cancer Patients, In Vitamin C and Cancer by A. Hoffer, Quarry Health, Quebec, Canada.

[531] Padayatty S.J. Levine M. (2000) Reevaluation of Ascorbate in Cancer Treatment: Emerging Evidence, Open Minds and Serendipity, Journal of the American College of Nutrition, Vol. 19, No. 4, 423-425

[532] Richards E. (1988) Vitamin C and Cancer – Medicine or Politics? London, England.

[533] Hoffer A. Pauling L. (1990) "Hardin Jones biostatistical analysis of mortality data for cohorts of cancer patients with a large fraction surviving at the termination of the study, and a comparison of survival times of cancer patients receiving large regular oral doses

of vitamin C and other nutrients with similar patients not receiving those doses, Journal of Orthomolecular Medicine, 5, 143-154.

[534] Hoffer A. Pauling L. (1993) Hardin Jones biostatistical analysis of mortality data for a second set of cohorts of cancer patients with a large fraction surviving at the termination of the study and a comparison of survival times of cancer patients receiving large regular oral doses of vitamin C and other nutrients with similar patients not receiving these doses, Journal of Orthomolecular Medicine, 8, 1547-1567.

[535] De Laurenzi V. Melino G. Savini I. Annicchiarico-Petruzzelli M. Finazzi-Agro A. Avigliano L. (1995) Cell death by oxidative stress and ascorbic acid regeneration in human neuroectodermal cell lines, Eur J Cancer, 31A(4), 463-466.

[536] Meadows G.G. Pierson H.F. Abdallah R.M. (1991) Ascorbate in the treatment of experimental transplanted melanoma, Am J Clin Nutr, 54(6 Suppl), 1284S-1291S.

[537] Riordan H.D. Jackson J.A. Schultz M. (1990) Case study: high-dose intravenous vitamin C in he treatment of a patient with adenocarcinoma of the kidney, J. Ortho Med, 5, 5-7.

[538] Riordan N. Jackson J.A. Riordan H.D. (1996) Intravenous vitamin C in a terminal cancer patient, J. Ortho Med, 11, 80-82.

[539] Riordan N.H. Riordan H.D. Meng X. Li Y. Jackson J.A. (1995) Intravenous ascorbate as a tumor cytotoxic chemotherapeutic agent, Med Hypotheses, 44, 207-213.

[540] Riordan H.D. Riordan N.H. Meng X. Zhong Z. Jackson J.A. Improved microplate fluorometer counting of viable tumor and normal cells, Anticancer Res., 927-932.

[541] Gonzalez M.J. Miranda-Massari J.R. Mora E.M. Jimenez I.Z. Matos M.I. Riordan H.D. Casciari J.J. Riordan N.H. Rodriguez M. Guzman A. (2002) Orthomolecular oncology: a mechanistic view of intravenous ascorbate's chemotherapeutic activity, P R Health Sci J, Mar, 21(1), 39-41.

[542] Raic-Malic S. Svedruzic D. Gazivoda T. Marunovic A. Hergold-Brundic A. Nagl A. Balzarini J. De Clercq E. Mintas M. (200) Synthesis and antitumor activities of novel pyrimidine derivatives of 2, 3-O, O-dibenzyl-6-deoxy-L-ascorbic acid and 4, 5-didehydro-5, 6- dideoxy-L-ascorbic acid. J Med Chem, 43(25), 4806-4811.

[543] Tashiro F. Sugiyama A. Urano Y. Kochi M. (2002) Sodium 5, 6-benzylidene-L-ascorbate induces in vitro neuronal cell differentiation accompanying apoptosis and necrosis, Anticancer Res, 22(3), 1423-1431.

[544] Esposito E. Cervellati F. Menegatti E. Nastruzzi C. Cortesi R. (2002) Spray dried Eudragit microparticles as encapsulation devices for vitamin C, Int J Pharm, 21, 242(1-2), 329-334.

[545] Clement M.V. Ramalingam J. Long L.H. Halliwell B. (2001) The in vitro cytotoxicity of ascorbate depends on the culture medium used to perform the assay and involves hydrogen peroxide, Antioxid Redox Signal, 3(1), 157-163.

[546] Benade L, Howard T, Burk D. (1969) Synergistic killing of Ehrlich ascites carcinoma cells by ascorbate and 3-amino-1, 2, 4, -triazole, Oncology, 23, 33–43.

[547] Bram S Froussard P. Guichard M. Jasmin C. Augery Y. Sinoussi-Barre F. Wray W. (1980) Vitamin C preferential toxicity for malignant melanoma cells, Nature, 284, 629-631.

[548] Helgestad J. Pettersen R. Storm-Mathisen I. Schjerven L. Ulrich K. Smeland E.B. Egeland T. Sorskaard D. Brogger A. Hovig T. (1990) Characterization of a new malignant human T-cell line (PFI-285) sensitive to ascorbic acid, Eur J Haematol, 44, 9-17.

[549] Noto V. Taper H.S. Jiang Y.H. Janssens J. Bonte J. De Loecker W. (1989) Effects of sodium ascorbate (vitamin C) and 2-methyl-1, 4-naphthoquinone (vitamin K3) treatment on human tumor cell growth in vitro, Cancer, 63, 901-906.

[550] Park C.H. Aniare M. Savin M. A. Hoogstraten B. (1980) Growth suppression of human leukemic cells in vitro by L-ascorbic acid, Cancer Res, 40, 1062-1065.

[551] Yamafuji K. Nakamurar Y. Omura H. Soeda T. Gyotok K. (1971) Anti-tumor potency of ascorbic, dehydroascorbic, or 2, 3-diiketogulonic acid and their action on deoxyribonucleic acid, A Krebsforsch, 76, 1-7.

[552] Yagashita K. Takahashi N. Yamamoto H. Jinnouchi H. Hiyoshi S. Miyakawa T. (1976) Effects of tetraacetyl-bis-dehydroAA, a derivative of ascorbic acid, on Ehrlich cells and HeLa cells (human carcinoma cells), J Nutr Sci Vitaminol, 22, 419427.

[553] Zheng Q.S. Sun X.L. Wang C.H. (2002) Redifferentiation of human gastric cancer cells induced by ascorbic acid and sodium selenite, Biomed Environ Sci, 15(3), 223-232.

[554] Gilloteaux J. Jamison J.M. Arnold D. Ervin E. Eckroat L. Docherty J.J. Neal D. Summers J.L. (1998) Cancer cell necrosis by autoschizis: synergism of antitumor activity of vitamin C: vitamin K3 on human bladder carcinoma T24 cells, Scanning, Nov, 20(8), 564-575.

[555] Sakagami H. Satoh K. Hakeda Y. Kumegawa M. (2000) Apoptosis-inducing activity of vitamin C and vitamin K, Cell Mol Biol (Noisy-le-grand), 46(1), 129-143.

[556] Jamison J.M. Gilloteaux J. Taper H.S. Calderon P.B. Summers J.L. (2002) Autoschizis: a novel cell death, Biochem Pharmacol, 15, 63(10), 1773-1783.

[557] Gilloteaux J. Jamison J.M. Arnold D. Summers J.L. (2001) Autoschizis: another cell death for cancer cells induced by oxidative stress, Ital J Anat Embryol, 106(2 Suppl 1), 79-92.

[558] Okayasu H. Ishihara M. Satoh K. Sakagami H. (2001) Cytotoxic activity of vitamins K1, K2 and K3 against human oral tumor cell lines, Anticancer Res, 21(4A), 2387-2392.

[559] Verrax J. Cadrobbi J. Delvaux M. Jamison J.M. Gilloteaux J. Summers J.L. Taper H.S. Calderon P.B. (2003) The association of vitamins C and K3 kills cancer cells mainly by autoschizis, a novel form of cell death. Basis for their potential use as coadjuvants in anticancer therapy, Invited review, European Journal of Medicinal Chemistry, 38, 451-457.

[560] Calderon P.B. Cadrobbi J. Marques C. Hong-Ngoc N. Jamison J.M. Gilloteaux J. Summers J.L. Taper H.S. (2002) Potential therapeutic application of the association of vitamins C and k(3) in cancer treatment, Curr Med Chem, 9(24), 2271-2285.

[561] De Loecker W. Janssens J. Bonte J. Taper H.S. (1993) Effects of sodium ascorbate (vitamin C) and 2-methyl-1, 4-naphthoquinone (vitamin K3) treatment on human tumor cell growth in vitro II, Synergism with combined chemotherapy action, Anticancer Res, Jan-Feb, 13(1), 103-106.

[562] Gilloteaux J. Jamison J.M. Venugopal M. Giammar D. Summers J.L. (1995) Scanning electron microscopy and transmission electron microscopy aspects of synergistic

antitumor activity of vitamin C - vitamin K3 combinations against human prostatic carcinoma cells, Scanning Microsc, Mar, 9(1), 159-173.

[563] Venugopal M. Jamison J.M. Gilloteaux J. Koch J.A. Summers M. Giammar D. Sowick C. Summers J.L. (1996) Synergistic antitumor activity of vitamins C and K3 on human urologic tumor cell lines, Life Sci, 59(17), 1389-1400.

[564] Jamison M. Gilloteaux J. Venugopal M. Koch J.A. Sowick C. Shah R. Summers J.L. (1996) Flow cytometric and ultrastructural aspects of the synergistic antitumor activity of vitamin C-vitamin K3 combinations against human prostatic carcinoma cells, Tissue Cell, 28(6), 687-701.

[565] Noto V. Taper H.S. Jiang Y.H. Janssens J. Bonte J. De Loecker W. (1989) Effects of sodium ascorbate (vitamin C) and 2-methyl-1, 4-naphthoquinone (vitamin K3) treatment on human tumor cell growth in vitro, Synergism of combined vitamin C and K3 action, Cancer, 63(5), 901-906.

[566] Taper H.S. Jamison J.M. Gilloteaux J. Gwin C.A. Gordon T. Summers J.L. (2001) In vivo reactivation of DNases in implanted human prostate tumors after administration of a vitamin C/K(3) combination, J Histochem Cytochem, Jan, 49(1), 109-120.

[567] Gilloteaux J. Jamison J.M. Arnold D. Taper H.S. Summers J.L. (2001) Ultrastructural aspects of autoschizis: a new cancer cell death induced by the synergistic action of ascorbate/menadione on human bladder carcinoma cells, Ultrastruct Pathol, 25(3), 183-192.

[568] Taper HS, de Gerlache J, Lans M, Roberfroid M. (1987) Non-toxic potentiation of cancer chemotherapy by combined C and K3 vitamin pre-treatment, Int J Cancer, Oct 15, 40(4), 575-579.

[569] Taper H.S. Roberfroid M. (1992) Non-toxic sensitization of cancer chemotherapy by combined vitamin C and K3 pretreatment in a mouse tumor resistant to oncovin, Anticancer Res, 12(5), 1651-1654.

[570] Zhang W. Negoro T. Satoh K. Jiang Y. Hashimoto K. Kikuchi H. Nishikawa H. Miyata T. Yamamoto Y. Nakano K. Yasumoto E. Nakayachi T. Mineno K. Satoh T. Sakagami H. (2001) Synergistic cytotoxic action of vitamin C and vitamin K3, Anticancer Res, 21(5), 3439-3444.

[571] Jiang Y. Satoh K. Aratsu C. Kobayashi N. Unten S. Kakuta H. Kikuchi H. Nishikawa H. Ochiai K. Sakagami H. (2001) Combination effect of lignin F and natural products, Anticancer Res, Mar-Apr, 21(2A), 965-970.

[572] Casciari J.J. Riordan N.H. Schmidt T.L. Meng X.L. Jackson J.A. Riordan H.D. (2001) Cytotoxicity of ascorbate, lipoic acid, and other antioxidants in hollow fibre in vitro tumours, British Journal of Cancer, 84, 11, 1544-1550.

[573] Lee K.W. Lee H.J. Kang K.S. Lee C.Y. (2002) Preventive effects of vitamin C on carcinogenesis, The Lancet, 359, 9301.

[574] Upham B.L. Kang K.S. Cho H.Y. Trosko J.E. (1997) Hydrogen peroxide inhibits gap junctional intercellular communication in glutathione sufficient but not glutathione deficient cells, Carcinogenesis, 18, 37-42.

[575] Kalokerinos A. (1974) Every second child, Keats Publishing Inc, New Canaan, USA.